THE UNLIKELY ADVENTURES OF A GEOLOGIST

MAGIC TRAVELS

PATRICK MCLAREN

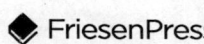 FriesenPress

One Printers Way
Altona, MB R0G 0B0
Canada

www.friesenpress.com

Copyright © 2022 by Patrick McLaren PhD
First Edition — 2022

All rights reserved.

No part of this publication may be reproduced in any form, or by any means, electronic or mechanical, including photocopying, recording, or any information browsing, storage, or retrieval system, without permission in writing from FriesenPress.

Cover photo of Mount Stephen, Yoho National Park, BC, and chapter line drawings by Susan Ferguson

ISBN
978-1-03-915101-7 (Hardcover)
978-1-03-915100-0 (Paperback)
978-1-03-915102-4 (eBook)

1. BIOGRAPHY & AUTOBIOGRAPHY, PERSONAL MEMOIRS

Distributed to the trade by The Ingram Book Company

For Susan, and my children Margot, Liam and Miles.

CONTENTS

Chapter 1:
Cambridge (1985)   1

Chapter 2:
Ottawa (1953-1963)   11

Chapter 3:
Jasper (1963)   26

Chapter 4:
Ottawa (1963-1964)   60

Chapter 5:
Jasper (1964)   73

Chapter 6:
Europe (1965)   97

Chapter 7:
Operation Bow-Athabasca (1966)   136

Chapter 8:
Baffin Island, Canadian Arctic (1967)   158

References   **207**

CHAPTER 1:
CAMBRIDGE (1985)

"What are the magic words?"

"Abracadabra!" chorused the voices of the twenty or so children grouped around my feet.

"Absolutely not," I said as solemnly as I could. "They have been changed this year to *gilly-gilly*. Can you say that?"

"Gilly-gilly," came the reply.

"Surely you can do better than that," I coaxed.

"GILLY-GILLY!" they shouted, bringing the roof down this time.

A cigarette appeared at my fingertips. Then another. And another. Four in all, three held between the fingers of my right hand. The fourth was between my lips, already lit with a Zippo lighter. "Of course, you should all know—never, *ever* smoke!"

I crushed the burning cigarette into my fist and opened my hand. My palm was unscathed, and the cigarette had disappeared. The children's eyes were as round as saucers. The rest of the show would take care of itself.

How many times had I said the same words and done the same simple tricks? Sometimes it had seemed so often that I felt almost like a charlatan. How could such trivial sleight of hand still fool even wide-eyed five- and six-year-olds?

Yet here I was, in 1985, performing them again, though this time after a gap of several years. The occasion was the birthday of the startlingly precocious son of a Cambridge professor, Donna's recently appointed PhD adviser. In an effort to be collegial, we had had him and his family to dinner at our upstairs Darwin College flat a few weeks before.

"Say hello to the McLarens, Benji." Dr. Lehman's voice had the full cultured fruitiness of a Cambridge don in an old-fashioned novel. He looked the part too, with his mop of unkempt hair and ill-fitting tweed jacket.

"Hello Benji," I said in a matching, tonsil-pinching tone.

Benji halted his ascent up the stairs. "I prefer Benjamin, Mr. McLaren. Some people call me Benji, but my real name is Benjamin. I would appreciate the courtesy of being addressed by my *real* name."

He proceeded up the stairs, squeezed past the adults who crowded the narrow hall, and jumped onto our bed, smearing muddy shoes on our prized Mennonite quilt. His parents beamed at him. Benji and I were bound to become fast friends.

Over the course of the evening I did some sleight of hand for Benji, starting with that perennial standby, the French Drop, making a coin disappear, then seemingly re-emerge from his ear. He rewarded me with some grudging admiration.

I confessed to his father that I had actually been a professional at one time, performing in nightclubs around the world, but had left my magic case in the basement of our house in Victoria when we had come to England. Later that night I had a sudden urge to see it again. We were low on money after the expense of moving to Cambridge. Perhaps I should try to do the odd show now and again. What could I charge? Twenty, thirty pounds for a half-hour? Even a couple of shows a week would help stop the drain on our meagre savings.

I went to bed that night with a nagging desire to get hold of the magic case. I didn't really *want* to do shows again, but there were a lot of memories in that battered suitcase. And it would be nice to perform the real deal again at least for friends. I had found myself showing off my sleight of hand skills to new friends and feeling frustrated not being able to follow up with some proper magic.

The next day I wrote our tenants in Victoria enclosing a cheque to cover mailing costs and explaining where the case was and what was in it. They would have to declare the contents to customs which made me feel a little foolish. A Cambridge Visiting Scholar (unpaid) who was supposed to be building a brilliant, high-paying career as a geological consultant was asking for his magic case to be sent over?

I had told everyone that part of my life was finished. I would have to swallow my pride, but the reality had to be faced. A year had passed without any contracts. Oil prices were down, a sure job with Mobil had fallen through, and the Canadian government, despite a number of promised contracts, had so far let me down. Our savings were disappearing at an alarming rate. Donna claimed that the key to success was to diversify and use all our available skills to get us through the hard times. Magic certainly fell into that category, so why not? Do it on the street corner in front of King's College Chapel if necessary.

The case arrived. It looked sad, worn, and shabby. The remnant of a sticker displayed a military coat of arms and the words *Alert, Inuit, Nunangata, Ungata*. With an odd feeling I realized I had never really read that before. It had been put on by a person unknown after my return to Alert in 1974 after a five-day, under-ice diving operation at the North Pole. A quick Internet search in 2020 revealed its meaning: *Beyond the Inuit Land*, the motto of the Canadian Forces

Station at Alert, the most northerly permanent habitation on the planet. I had performed a show there during one of that expedition's numerous parties.

Inside, the contents of the case looked old and pathetic. The box for the Disappearing Die was broken, the die itself falling apart. The case had had a rough journey. It seemed improbable that these motley bits of cards, rope, and handkerchiefs would be enough to put on a whole show. Tobacco crumbs had gathered in the nooks and crannies of the lining. A little deflated, I shoved the case on top of a dresser, wondering why I had spent seventy-five precious dollars to get it here.

That evening the phone rang. It was Dr. L. Could I possibly do a show for darling Benji? His sixth birthday was next week, and he so wanted to see some more magic.

"Well," I said cunningly, "I don't come cheap."

"No, no, of course not." An ingratiating chuckle. "What if you bring Margot and Donna and stay for some tea?" He made it sound like an invitation to Buckingham Palace would be a miserable gesture compared to this. Donna's future PhD was in my hands.

"I'd be delighted," I said and then covered the mouthpiece with my hand. "Your goddammed supervisor has just coerced me into doing a show—for free."

Work had to be done. Poster paints were bought, and somewhat reluctantly, I let Margot paint new spots on both the real and false dice. A nine-year-old can be most insistent, and she had long considered Daddy incapable of doing anything arts-and-craftsy. I borrowed tools from Clare, who lived in the flat next door and had just turned ninety years old. At seventy, Clare had learned Medieval Latin. At eighty-eight, her first book had been published, a history of a little village outside Ely. It had become a best seller, for a scholarly text anyway, and now her services as a translator of Medieval Latin manuscripts were sought by historians and academics across the land.

We had met her shortly after moving in. Tiny and bright-eyed, she exuded a *joie de vivre* that was irresistible. She had listened carefully to our names, and several weeks later, the doorbell rang. It was Clare.

"Come in, come in," I said. She bounded up the stairs, an odd look on her face, then she handed me two letters, unstamped but addressed in my parents' handwriting.

"Now where do you suppose I got these?" asked Clare. I imagined several ways, but all seemed impossible. "Yesterday, I had tea with your mother in Ottawa," she announced with great satisfaction. I didn't hide my astonishment, but the explanation was simple enough.

Clare's sister was the mother of friends of my parents, all of whom lived in Ottawa. Because her sister was rather getting on a bit, Clare had gone over to Canada for a visit. She had often seen my mother on previous trips, and when my mother had mentioned casually that her son Patrick had just moved to Cambridge, Clare thought for a bit. "I believe he has moved next door to me!" So the mail got personally delivered.

Clare came in handy in another equally surprising way. For some reason, she possessed a smaller version of the clippers used to trim horses' hooves and armed with these and a large piece of local flint as a hammer I was able to mend the false door of the die box. With Margot's help, all the other tricks were renovated and the case cleaned and polished. I finished by making a complete list of all the tricks needed for a half-hour show as well as those used in close-up work and taped this to the inside of the lid for reference. Cigarette sleights, disappearing thimble, rope cutting, linking rings.... The list didn't look like much, but they were all old faithfuls. I used to pride myself (or was it to reassure myself?) that my magic was the simple, straightforward kind. The pleasure was in the simplicity and the sleight of hand. No one could explain away an effect as simply apparatus. There is no surer way to belittle the skill of a magician, but my only real apparatus trick was the Disappearing Die. Although nearly all required at least some small unseen gimmick, the magic effect always depended on two basic skills, manual dexterity and misdirection. Of these, misdirection had been by far the most difficult to acquire.

Misdirection is the art of drawing attention to where it's wanted and away from the crucial moves of the sleight. Take the French Drop for example. Although not really a trick that works all on its own (except with small children), it may be required in many close-up effects. It is a simple, well-known manipulation in which it appears as though something, usually a coin, is taken from one hand into the other and then disappears. The manipulation occurs the instant the coin is momentarily covered by the taking hand, when it is allowed to drop unseen into the palm of the hand holding it in the first place. The misdirection comes in the follow-through. The hand that supposedly contains the coin moves in one continuous arc away from the body with the magician's eyes (and consequently the audience's) following it. If the magician really believes the coin is there, the audience will too. A beginner will try too hard to make the guilty hand as inconspicuous as possible, thereby announcing the false move in body language equivalent to a loudspeaker. An experienced magician will have long since forgotten about the hand that really holds the coin. The important thing is what to do with the coin when it has been taken. The guilty hand has become irrelevant.

Misdirection has to become an instinctive part of the body language of the magician. Every move must look natural; if it doesn't, it will arouse suspicion. That's why some of my tricks, no matter how expert my manipulations became, had never been successful. The Multiplying Billiard Balls for example consistently failed to impress. Despite many hours in front of a mirror practising near-perfect manipulations, my hand movements never managed to look like part of my ordinary everyday movements. The body language was simply wrong. It's not good enough to just fool the audience.

An effect has failed when spectators suspect a false move even if they don't know what it might have been.

An afternoon's work got everything in the case back in working order, but apart from the sleights, I hadn't done any of these tricks for the better part of six years. When I went through them all for Margot and some of her school friends though, I found that my hands had a memory of their own. It was almost a sensuous pleasure to feel the slip of the three uneven ropes becoming equal or the way the linking rings clanged as they irresistibly joined each other to form an interlocking chain. Occasionally, my memory failed me, and I had to go back a few steps to allow instinct to carry me through. Even the patter came flooding back. Always the same words at the same moments. I had never been showman enough to attempt a witty or comedic act, but I had found a successful formula in appearing somewhat diffident, a little uncertain that my tricks would work, then relieved and thankful, even a little surprised, when they did, as if taking pleasure from the fact that the effect had actually fooled my audience. In this way, I found I could quite easily manipulate my viewers. They naturally want to find out how a trick is done. At the same time, most don't want to be embarrassed by an inept entertainer. So they are also relieved when the magic works and the magician has not made a fool of himself, and they can ask, "Seriously though, how did you do that?" It both amused and annoyed me when I did close-up table magic in nightclubs and restaurants and was told, almost as a confession, "I really couldn't see how you did that!" I don't think a magician would last long otherwise.

But did I really want to do magic shows regularly again? It had only been marginally fun to treat Benji and his friends to a surprise birthday show. My magic was really more suited for adults than children. I had never bothered with all the paraphernalia required for the kind that kids enjoy with lots of colourful props and endless productions of silk scarves and paper whirlies. Children are not as impressed as adults when their watches disappear. Chances are they are not even wearing watches. Above all, a rabbit is mandatory equipment for a kids' show, and my limited experience with rabbits had not been pleasant. More than once, my rabbit-producing apparatus had bounced itself across the table with angry thumping clearly audible from its interior. Rabbits also need to be lucky to survive the maulings of a roomful of excited children. I used to give them away to the lucky birthday child until I caught sight of one mother's look.

No. Deep down, I didn't really want to become a performer again. But we needed the income, needed it urgently, and it seemed crazy to be fussy when desperate. Still, *doing* magic wasn't the only income-generating scheme available. We had met Alex and Michael Allerhand shortly after arriving in Cambridge. Having married early, this ideally suited pair, enabled by a small inheritance, had settled down to the strenuous life of 1960s hippie dropouts. They purchased a broken-down cabin in the French Pyrenees and had lived there largely on potatoes, which were plentiful as well as cheap. No crass emotions such as ambition marred their tranquil existence until, one day, a bombshell hit. Michael received a curt instruction from Her Majesty's Government that it was essential to obtain a job or lose all his welfare payments. What could Michael do? He had never worked a day in his life, but there was one way out of the dilemma: enroll in a government-funded training program. With characteristic fervour, he selected the course with the longest waiting list. As long as he was on the list, the dole would continue.

Life went back to normal. Several placid years passed and then, to his surprise, he was ordered to attend a radio-electronics adult education program. His number was up. With no choice in the matter, Michael took the course and, in another surprise, excelled. At the end, his instructor told him he should go to university. Michael had never bothered to get any A-levels, but now he completed them in six months and then proceeded to take a bachelor's degree with grades sufficient to get into a graduate program in Engineering at Cambridge. When we met, he was just finishing, in record time, his PhD on computer speech recognition. Offers to publish the thesis had already rolled in, and the ex-hippie had just been made a fellow of Darwin College. Meanwhile, Alex also completed her A-levels, was accepted into Cambridge, and was about to finish a degree in history. Their success story is a tribute to the British adult education system.

Today in 2021, equipped with my iPhone, I am dependent on its Speech Interpretation and Recognition Interface known as Siri. Provided with an attractive female voice, Siri, according to Apple is also a Norse name meaning *a beautiful woman who leads you to victory*. In my mind, I bow in acknowledgement to Michael each time I strike up one of my all-too-frequent conversations with her.

The Allerhands lived near us in another Darwin College flat. Margot played with their two daughters, Rhalou and Taffeta, and I often amused Alex and Michael with stories from my days as a magician. One day, Alex suggested that there was a potential book in the magic anecdotes. She came from a literary background, her mother being a well-known children's author. The thought of writing a book was not new to me, only the subject. In Cambridge, everyone writes books. It is expected. In fact, on the salaries professors get, it is essential. I had been thinking along the lines of *The Coastlines of Canada: An In-depth Look at the Morphology, Geology, and Evolution of the Canadian Shoreline*. I could picture the tome in my mind's eye. A glossy coffee-table book explaining the origins of the top ten beaches of Canada and full of spectacular pictures. No Canadian household could be without it.

But Alex's idea had more immediate appeal. Maybe this was a better way to turn magic to advantage but without having to do shows again. Much has been accomplished in times of financial need. Hadn't Handel written the Messiah in twenty-four days to save him from starvation? The main hurdle, I decided, was to overcome the natural reticence of writing about oneself. But I had enjoyed many memoirs without feeling that the author was a pompous egotist. What about Miles Smeeton and his sailing books? Or Farley Mowat and his *Boat Who Wouldn't Float*, which had always struck me as a pack of lies until I actually had to charter a Newfoundland vessel to do geology in the Arctic. After nearly sinking in the Labrador Sea, I knew that everything he wrote had been the unexaggerated gospel truth. So to quote from my first writing attempts in Cambridge in 1985, "while I wait for Sweeney Todd's pizza parlour to line me up with magic shows for their Saturday afternoon birthday parties or for Shell to give me the go-ahead on a dive survey of their latest proposed pipeline route in the Indian Ocean, I shall try to get started."

CHAPTER 2:
OTTAWA (1953-1963)

I remember precisely when my interest in magic started. Christmas 1953. I was six years old. The occasion was the annual staff children's party at the Geological Survey of Canada. At that time, the Survey was housed in one wing of the National Museum in Ottawa. The party took place in the museum's auditorium, where many years later in December 2004, I was to read my father's eulogy to an international crowd of geologists. At the party, a red light on a huge map of Canada kept track of Santa's latest position on his route from the North Pole, steadily heightening our excitement as it moved across the map. But each new position was not always closer to Ottawa. At such moments, a pretend radio operator on the stage would hurriedly contact Santa for more details. There were always excruciating delays. A snowstorm over Chesterfield Inlet forced a detour eastward to Ungava Bay. To keep us from getting too frantic, quizzes and contests helped us forget Santa's plight until someone would shout out a new position on the map. Radio contact would be re-established but the news was never good. At one point, we learned that one of the reindeer had broken its leg while landing on a rough bit of tundra.

This was when the magician appeared.

Cutting a dashing figure in a tuxedo and top hat, he dazzled us. Our wonderment was absolute. How was this possible? Each effect was more stupendous, more thrilling, more elaborate, and more baffling than the last. The climax came when he explained that he had had an accident with his bunny and produced a limp paper cartoon rabbit from his back pocket. He was visibly upset and sorry that he had sat on poor little Jeremy on the way over. Then, to our alarm, he placed the cut-out in a pan and put a match to it. The flames quickly got out of control. He clamped a lid on the pan and with a piercing glance at us attached a bicycle pump to it. A few swift strokes and the lid was removed. Voila! A real live rabbit filled the pan.

It wasn't so much that I wanted to know how the magic was done. I wanted to be able to do it. I began to get books on magic from the library. Anything I could get my hands on was carefully studied. Luckily, I had a fellow enthusiast in my friend David Defries. Together we conducted what seems in retrospect innumerable magic experiments, taking turns to invent, no matter how outrageous, what we imagined were unprecedented new effects. I can't remember producing a successful one. On one occasion, it seemed a good idea to tape our *drop*—a hidden receptacle for whatever needs to be secreted, usually in the form of a small pocket overhanging the far side of the magician's table—onto a pant leg. This bit of brilliance required the whole show to be performed with only one side of the performer's body visible to the audience. Not exactly a natural look.

There was, of course, a limit to what we could learn from the kinds of simple and often impractical books that were available at the library, even with all our creative variations. I remember bowls of rice, glasses of coloured water, and a succession of fishy-looking homemade tubes in one strange arrangement after another. Every book emphasized that a magic show had to flow easily from one trick to another, yet our attempts all seemed to require elaborate preparation prior to every trick. It may be all right to send a skeptical sister from the room between each effect, but you can hardly ask an audience to close its eyes for a few moments whenever another set-up is needed. Also, and there was no getting around it, most of the tricks stretched credulity to the breaking point. It was simply not logical to show one tube empty, put it carefully over a second tube, then mutter something about having forgotten to show how empty the second tube was and slide them apart (carefully) to prove the emptiness of the second before putting them together again, and ta-da! Silk scarves from nowhere. No matter how hard we worked, we never seemed to attain the happy state depicted in the numerous illustrations of a young magician confidently performing before a rapturous group of, we presumed, paying customers.

Our coup de théâtre was a huge, unwieldly piece of apparatus consisting of a six-foot-high plywood board with a seat on either side. Volunteers were invited to examine it at close quarters, but not for long. The workmanship was not of the highest quality. David and I sat on each side, and a curtain was drawn, hiding both us and the apparatus. Grunts and grating sounds followed, accompanied by a violent shaking of the board. The curtains were whisked aside. We had changed places!

Our biggest difficulty was finding an audience. Cynical siblings are unsatisfactory. To get better, we needed to perform for someone who hadn't seen the trick before. Whenever such a person was found, one of three things would usually happen. If the trick failed (meaning the spectator saw how it was done), scorn and derision would be heaped upon us, and the (now former) friend would have the satisfaction of feeling clever. When a trick worked, the situation was reversed. We were now the clever ones and, feeling good about ourselves, would usually be persuaded to repeat the performance.

Unfortunately, there is an overwhelming tendency to keep showing off a successful effect until its secret is figured out. On those rare occasions when enlightenment never arrived, the urge to heighten our prestige by explaining the trick was usually irresistible. A complete loss of interest inevitably followed.

Summoning all our courage, we advertised in *The Neighbourly Natter*. The plan was to alternate so that while one of us performed the other could prepare the next trick. As an added benefit, the other could start clapping should the bafflement we inflicted on our audience cause them to forget this small courtesy.

Our big break came when Bill's Joke Shop got a new assistant. Situated on a rather sleazy section of Bank Street, a block away from where the nicer shops started, Bill's did a thriving trade in plastic dog turds and other, mainly scatological, humour. One bestseller was a small rubber pouch that fastened beneath the lid of a toilet. This was filled with warm water and placed with its nozzle directed upward. For the illiterate, or possibly to comply with Canada's burgeoning bilingualism, a sequence of cartoon illustrations showed the set-up procedure. I had no interest in such stuff, but I spent a lot of time at Bill's for another reason. Half-hidden among the plastic vomits and gifts for nymphomaniacal secretaries were a small number of packaged magic tricks. These included miniature versions of the linking rings, the cups and balls, and false thumbs.

Bill himself would do magic when pressed, but there was no money in it, and he concentrated his efforts on keeping an eye on his often rowdy and light-fingered clientele. I was a little frightened of Bill. He was squat and toad-like with bulgy eyes behind coke-bottle glasses. His considerable temper was never far below the surface. I thought him unreasonable for refusing to show me the secret of a trick before I had bought it. After one such altercation, he ordered me out of his shop. Still, I have one important thing to be grateful to Bill for. It was he who sold me the Disappearing Die. Perhaps sensing that I would I find it irresistible, he produced it one day from his inner sanctum behind a grubby curtain just beyond the displays.

I stared in awe at an exquisitely made rectangular box with four hinged doors, two in the front and two on top, with a partition between the two halves. It was a big cut above the junk he sold in the store and clearly wasn't new. I never found out where it came from, but it was beautiful. One of the front doors concealed a false panel painted to look like the face of a die. A hidden mechanism in the doorknob released or retrieved the panel. There was also a built-in slide that produced a noise when the box was tilted, making the audience sure that the die had simply slid over to the other side. This refinement heightens the sucker effect. After the performer appears to place a solid die into one half of the box, the false panel is displayed, removing any doubt that the die is really there. All the doors are then closed, and the box is tilted sharply, activating the sound of the slide. The door to the chamber originally containing the die is flung open and...where has it gone?

The audience would be an unusual one if some Clever Dick failed to shout out the obvious solution, "It's in the other side!" That is the magician's cue to close the door again, give the box another ostentatious shake, perhaps with a slightly embarrassed guilty look while doing so, and then open door on the other side. It too is empty. It doesn't take long before Clever Dick's triumphant shouts of "Open all the doors!" fill the room. Careful judgment is required to determine the optimum moment to comply, usually in the middle of the loudest protestations. Total silence. The die has really gone! The magician knew what he was doing after all. CD is an idiot, and the rest of the audience is left feeling like suckers for being influenced even for a moment by him. But the magician is magnanimous. No hard feelings, folks. The die is really over here in this previously empty hat.

It was a trick I loved. It was foolproof, and the only completely self-working piece of apparatus I have ever used. I had already constructed several cardboard versions of it before Bill showed me the real thing. God, how I wanted it! It was the added touch of genius in the built-in noisemaker that overwhelmed me. Bill wanted thirty dollars for it. I whinged and whined for all I was worth. He lowered the price to twenty. It was a fair price, all in all, but it took many weeks of shovelling snow and delivering newspapers before I became the proud owner.

One day, David and I were doing our rounds of Bill's latest equipment when we saw a new person behind the counter. He had on a worn suit that matched his sallow complexion. His appearance was disreputable, and his eyes darted restlessly about, as if constantly searching for real or imagined dangers. I was destined to become closely acquainted with Ed Denis, but I never saw him without that hunted look. A group of kids stood before him, and as I watched, I felt a tingling shoot down my spine. I was witnessing real magic for the first time. Ed's bony hands and wrists emerged from frayed, soiled cuffs, his nicotine-stained fingers capped by uncut, grubby nails. He held a lit cigarette between each of his fingers, and smoke curled from another between his lips. David and I stood at the back transfixed with astonishment as cigarettes effortlessly appeared and disappeared. One moment a cigarette was between his fingers, the next it was gone. Magically, it rose again from his closed fist only to disappear once more, this time reappearing from behind his elbow. Nothing we had read in all our books of magic could explain it. An elastic pull? Impossible. A drop? The counter was glass. Yet from empty hands a cigarette would appear, then another and another. A lit cigarette went into his fist and promptly disappeared. We stood there frozen, torn between the sheer wonder of it and frustration at the seeming impossibility of ever learning such *real* magic.

Magic Travels

Perhaps we believed it was real magic in another sense. Since neither of us could conceive of how such effects were done, real magic seemed the only alternative. We left the shop without speaking to him, but we could talk of nothing else with each other. We walked down Bank Street shivering with excitement. "Did you see how...?" "It couldn't have been a pull." "How do you suppose...?" Seeing that performance by Edmund Denis was the turning point for me in understanding what magic could really do.

It's curious, but I don't remember when I first summoned up the nerve to speak to Ed. I was a shy kid, so it was probably a gradual process. But Ed was my first experience of what I later found with all the real magicians I ever met. They are unfailingly a friendly, hospitable lot, always ready to encourage a genuine enthusiast. Ed took me on. Slowly, painfully, I learned the secrets of proper sleight of hand. Fortunately (for me), my mother smoked, and there was always a ready supply of cigarettes in the house. She kept her Players Navy Cut (no filters) in a silver box on the living room coffee table. In short order, I became an expert on the brands most suitable for sleights. Unfiltered Players Navy Cuts were excellent. Filters and king size were better avoided. For the duration of my early teens, my mother seldom selected a cigarette from her silver box that wasn't bent or mangled or missing half its tobacco. It was impossible for me to enter the living room without filling my palm with cigarettes and manipulating them in front of the mantelpiece mirror. I did not confine myself to cigarettes. Multiplying billiard balls were the bane of my existence. I liked practising with them during homework sessions in my bedroom. The noise of balls rolling across the floor told my parents that the work going on upstairs was not entirely scholastic.

With Ed's tutoring, the descriptions of tricks in books took on new clarity. Learning magic from books is a frustrating business. Even the simplest effect takes many pages to explain. The impact often gets lost in the sheer number of words needed to describe it. Thanks to Ed, I began to understand the principles better and appreciate the subtleties of misdirection that were required to change a trick from *not another card trick* into something beautiful. Ed himself used to put on the occasional show. I learned much by helping him backstage with the props. His costume and stage persona were vaguely Chinese in a Charlie Chan sense, and I have used his magic words, *gilly-gilly*, ever since.

∞

Much of my practising relied on the goodwill of school friends. The school was Glebe Collegiate, an imposing three-storey building that held some 1,500 of us in Ottawa's solidly middle-class Glebe District. We were told it was a good school at the time, and in hindsight, it probably was. Among a number of classmates who became reasonably well-known alumni, the most famous was probably Luba Goy, a gifted comedian who rose to fame as a member of the Royal Canadian Air Farce. Whether in class or in the school's auditorium, her humour was brilliant. School was never a bore when Luba was around. My own circle of friends included John, slight and cool, who smoked secretly and eventually became head boy and then surprised me by revealing that his sole purpose in life had been to achieve that lauded position. Mike, the son of a United Church minister, was a kind, gentle soul whose religious upbringing came into focus when I visited his house and heard his mother scolding his younger brother for saying *darn!* when a Dinky toy broke. When it was discovered that Mike's father had been naughty with not one, but many, female members of his flock, the scandal rocked the neighbourhood. Don Hindle, together with John and Mike, made the third member of a folk group known as The Terriers. With significant similarities to the Kingston Trio, they were a staple at school assemblies. Don was as slight as John and popular too. He had a unique sense of humour that I found myself emulating. Even now, to get a laugh, I use self-mocking phrases such as "People forget that I've got feelings too, you know" that I'm pretty sure originated with Don. His back was marred by a long and impressive scar where seven vertebrae had been fused to repair a spinal curvature. It had meant long months in a body cast as a child.

Two others made up this intimate core of buddies. Keith Ogilvie's main claim to fame at the time was his highly attractive sister Jeanie, who in later years was, for a while, Don's wife. Keith was exceptionally bright, and his family was the most welcoming I've ever known. Anyone with the good fortune to enter their house was met with incredible warmth and hospitality. His father, Skeets, had been a decorated fighter pilot in the war. Famous for his role in shooting down a German Dornier over Buckingham Palace just before it dropped its bombs, he was later hit and badly wounded in aerial combat over Europe. Captured and hospitalized for months, he ended up in the notorious Stalag Luft III POW camp. There he was instrumental in the intricate planning for the Great Escape made famous by the 1963 Steve McQueen movie. Skeets was the last person through the tunnel before a patrol guard, who had paused by the tunnel's exit to urinate, spied the next escapee emerging from the hole.

The escapees had been made to understand that crossing Hitler's new autobahn just beyond the camp would be highly dangerous, so Skeets had waited three days, hidden in the brush, before he found the courage to cross. It was only a few days later, after hiding in woods and swamp in brutally cold winter conditions, that he was forced like so many of the other escapees to emerge onto roads where German soldiers, not even bothering to look for them in the bush, were waiting to pick them up. It was, being still alive, the greatest relief when he was recaptured. After being kept in a variety of jails and interrogated relentlessly, he was eventually taken back to

Stalag Luft III. There, he learned that of the seventy-six men who had escaped, fifty, randomly selected during the roundups, had been executed by the Gestapo. Hitler had ordered all to be executed but was tactfully persuaded by Göring that this would too obviously be a crime against humanity. The Führer then insisted that more than half must be shot. None of the attending officers dared to contradict him this time. It was only after liberation when Skeets discovered that three had made it to safety.

We all shared a deep respect and affection for Skeets. He was never anything but humble about his war exploits, and it was a pleasure to listen to him when he sometimes opened up even a little. With a notably square jaw, clearly depicted in the many sketches and caricatures by his fellow prisoners on the walls of various rooms, he invariably pulled on his pipe whenever he described the realities of the Great Escape. Without directly criticizing the movie, a great hit in its day, Skeets would gently suggest that Hollywood had omitted a few details and put just a little more emphasis on American heroism than was warranted. In reality, only one American prisoner participated in the planning, and the considerable Canadian input was ignored. Keith later wrote *The Spitfire Luck of Skeets Ogilvie: From the Battle of Britain to the Great Escape*. It was published in 2017 and is an outstanding tribute to his parents.

Guy Sprung was a late arrival in our third year at Glebe. Tall and good-looking, he aroused great curiosity when he showed up one day in class. We soon realized that he seemed to be good at everything and, it has to be admitted, didn't mind making that clear to anyone who cared to listen. Sports, science, math, English, and music all seemed to come effortlessly to him. His mother was German, and Guy had grown up speaking German and French, later adding Russian to his repertoire when he directed *A Midsummer Night's Dream* in Moscow. His Canadian father, Colonel Spike Sprung, was a military historian who later became a philosophy professor.

When we met him, he was learning Sanskrit and afterward produced numerous scholarly articles on Buddhism and the nature of being. Guy and I played flute duets together and somehow ended up in a competition where we played a Mozart piece and came second. The honour was only slightly marred when we learned we had been the only contestants in our category.

Among our many other achievements together was a hot air balloon that we perfected one evening in Patterson's Creek Park near Guy's house. It consisted of a large plastic laundry bag with aluminum foil taped around it to form a skirt at the bottom. The width of the skirt proved critical: too wide and the extra weight kept the balloon firmly on the ground, too narrow and our heat source, a flaming, alcohol-soaked ball of cotton wool that hung on wires, melted the plastic. The length of the wires also had to be factored in. To our great surprise, that night our ill-considered plan worked.

It was already dark when, after countless design alterations and failed attempts, our balloon unexpectedly filled up without a hitch and soared off into the void. It quickly became invisible save for a small flaming yellow ball. That was when the implications of what we were doing finally dawned on us. We were horror-struck! A stiff wind was blowing, and the ball was getting smaller fast. In a panic, we sprinted after it, terrified that we were soon going to be responsible for the Great Fire of Ottawa. How long would the fireball last? It was impossible to say. Where would it land? We sure as hell had to find out. I could already picture myself banging on the door of a stranger's house shouting, "Your roof is on fire! Your roof is on fire!" One thing was certain; we must not lose sight of our burning balloon. Heading southwest, we sprinted across Clemow Avenue and an exceptionally busy Bank Street. We couldn't stop for cars. With the palms of our outstretched hands imploring oncoming traffic to miss us, we kept our eyes skyward. I was only dimly aware of screeching tires and shouted curses. But we dared not stop. By the time we reached First Avenue, the ball of fire was descending. Lower and lower it came. Breathlessly, we watched it gently come to a halt, still burning, on the roof of what was then Carleton College. There was not a moment to lose. We found an open door and ran up four floors to the rooftop. The building was deserted. How we got onto the roof is lost to memory, but onto the roof we got, scrambling over the curved interlocking tiles just in time to watch the last of the dying sparks, leaving us alone in the darkness.

∞

These were the friends who saw me through the early stages of magic—often mocking my attempts to astound them, feigning boredom when confronted by yet another new effect, and even, if asked, providing dubious help to build my increasingly ambitious apparatus. By now, working with Ed had made me believe I could put on an equivalent spectacle for my high-school assembly. The showpiece was to be the Chinese Triangle Magic Screen. We constructed three cardboard panels, decorated on both sides with what we thought were appropriate Chinese symbols and hinged together to form a continuous screen that could be folded to make a standing triangle. Each panel was supported by a pair of legs that enabled the structure to stand on its own. The magician first shows one side of all three panels, which are then put together, momentarily forming the triangle in order to turn the screen around, which on reopening, displays the reverse side of the panels to the audience. The triangle is then reformed, and with the appropriate patter, the magician reaches into its centre cavity and produces an amazing variety of stuff, usually innumerable silk scarves, flags, and any other paraphernalia that can make an entertaining effect.

Through Ed, I had access to a large, evil-looking boa constrictor named Buttercup. I decided the reptile would create a spectacular effect after the pretty, but trivial, production of silk scarves. Now to understand fully the missteps that led to disaster, the mechanics of the screen have to be explained. A rectangular piece is cut out of the middle panel and replaced by a hinged bin that can flip back and forth from one side of the screen to the other. With the three panels of the screen facing the audience, the bin is positioned to hang on the unseen side. The front of the bin is camouflaged with the rest of the panel by the Chinese decorations, making the seam invisible. As the screen is folded outwards to form the triangle, the bin is pushed through to the other side with a knee, and protected from exposure by the enclosing, outside panels. The apparatus is then turned around and opened up to display the reverse side of the three panels. Once again, the bin hangs off the side away from view. The screen is then remade into a triangle with the bin now hidden inside ready for its contents to be miraculously revealed from out of nowhere.

The problem was in the construction of the bin. Made principally of cardboard and tape, it wasn't strong. Flimsy might be the appropriate word. The size of the bin was orders of magnitude larger than typical concealment props, promising, it seemed to us, a limitless capacity to produce startling effects. Our critical thinking was overwhelmed by our enthusiasm. I filled it to capacity. For the finale and pièce de resistance, I coiled Buttercup snugly into a compartment at the bottom. But I failed to take the snake's weight into account. At about seven feet long and thicker than my arm, Buttercup, though placid in temperament and an experienced performer, weighed upward of thirty pounds.

The show had gone well. The school's cheerleaders, who had performed ahead of me and were gratifyingly enthusiastic, now sat in the front row of the packed auditorium. I had just completed the linking rings with no shortage of excited and scantily clad assistants available for the choosing. All that remained was the Chinese Triangle Magic Screen which had been standing on display at one side of the stage. I was eagerly looking forward to what I knew would be a fabulous finish when the full length of Buttercup emerged out of the screen and wrapped his writhing body around my arms and shoulders. I could already anticipate the thrilling screams of delight and horror from the front-row girls. I lugged the screen to centre stage, opened and displayed the three front panels, then formed the triangle, surreptitiously pushing the bin to the other side with my knee. CRASH! Buttercup's mass tore the hinge off the panel, the bin collapsed, and the huge snake was hurled out across the floor of the stage. This was too much for Buttercup's even temper. In a panic, he shot out toward the cheerleaders in the front row and froze at the edge of the stage only a foot or two from their terror-stricken faces. Unsure what to do next, Buttercup raised his head, opened his jaws extremely wide, and with his ghastly tongue darting rapidly in and out released a series of horrifying hissing noises. It was not the ending I had envisioned. To be sure, there were plenty of screams. In fact, there was panic. The terrified cheerleaders stampeded for the exits, followed by everyone else. The principal, who had been emceeing the proceedings, shouted for calm and was ignored. I ordered the curtain lowered and rescued Buttercup, by now hiding inside one the footlight wells but none the worse for wear.

The debacle, which I didn't live down for the rest of my high-school career, was an especially dramatic example of the hard lessons that have to be learned to become a competent magician. It was this episode that convinced me that simple was best, at least for me. From that point on, I avoided any trick that required an elaborate set-up or that might invite the smart alecks to declare, "It's just the apparatus; anyone could do that with the right equipment." I went to great lengths to eliminate all obvious props. I was once accused of hiding cigarettes in my watchband. Since then, I have removed it before performing. And I always rolled up my sleeves. My only essential garment was a plain suede-leather vest that concealed a pull, an essential piece of magician's equipment that is used to disappear silk handkerchiefs. I also wanted to be able to travel easily, so my whole show had to be contained in one small case. At this stage, I never thought of making magic anything more than a hobby, but my teenage years were spent largely dreaming of travel. Doing magic along the way, I reckoned, would be a perfect way of making friends in strange places.

CHAPTER 3:
JASPER (1963)

"HEEEEEEEAAAAAGH! Come on Popeye, you goddam sonofabitch big-ape!" Popeye had come to a stop at the head of a fully packed line of horses on a dangerously narrow trail that hugged the mountainside and was far too narrow for any others to get past him. My hand reached behind into the saddlebag I had thoughtfully filled with stones and yanked one out. Rising up in my stirrups, I threw it as hard as I could over the single file of horses contentedly blocking any further movement. Never being much good at throwing, the stone went left, missing the target but dislodging a small avalanche of scree, which tumbled down around Popeye's ankles. He slowly swung his head around and looked at me with his usual slightly cockeyed expression. It was sympathetic and reproachful at the same time. "It's no good hurrying me. You know this is my favourite spot," he seemed to say. I was used to him stopping at this point and had come to believe that Popeye genuinely enjoyed a good view.

And the view from here was one of the best. To my left rose the mighty flank of Mount Edith Cavell in Jasper National Park. To my right it dropped away into a maze of scrub brush, then, far below, densely packed spruce and pine. Ahead lay the Astoria River Valley surrounded on either side by the full splendour of the Rocky Mountains. Even coming into Jasper for the first time by train, I couldn't suppress my excitement at seeing them. Sure, I had seen lots of pictures. I had even been in the mountains before as a young child. But nothing had prepared me for this. No doubt to the annoyance of my fellow passengers, I had darted from one side of the carriage to the other, afraid to miss any one of the spectacular sights as they progressively unfolded. What would it be like, I wondered, to make my way up that gully, cross over to the shoulder silhouetted by the western sun, then up into the snow until I stood at the top? The very top! The thought made me feel wild. I wanted to fly up to the heights, see them close up, touch them. It made me want to laugh with joy. How terrible to be an earth-bound boy.

The train pulled into Jasper station to welcoming shouts from the crowds that lined the platform. I eagerly scanned the faces in the hope of being recognized. My fellow passengers were already jamming the platform. They swarmed around the souvenir stands that sold genuine Indian crafts, but in half an hour, most would be back on the train heading west, having done Jasper. I shouldered my duffle bag and pushed my way through the throng to the street. I stared with interest at the grave, brown faces of my first real Indians. The air smelled sweet—mountain air sharp with the scent of conifers. The mountains stood still now. I could drink them in. I may have gaped like a simpleton. To my amazement, others seemed indifferent to the magnificence all around them.

It was a few moments before I was able to gather my wits and take stock of where I was. Across the street, I spied a sign in a drugstore window – TOURISTS WELCOME.

"Pardon me, ma'am," I ventured shyly to the large woman behind the counter. "Would you have any idea where Tom McCready might live?"

"Tom McCready?" she said, peering out from among her miniature canoes and plastic Hiawathas. "I thought everyone knew Tom." Then, noticing the duffle bag at my feet, she smiled. "Are you his new hired hand from the east?"

"Yes. He was supposed to meet me at the station." How marvellous that the first person I asked already knew who I was.

"Go out the door and turn left, then left again at the first street you come to, and go down about six or seven blocks 'til you reach Miette Street, M-I-E-T-T-E. You'll see a little brown bungalow on the right. Number 31. That's Tom's place."

It was further than the woman had made it sound. Away from the main street, I worked my way along a broad avenue with a few scraggly trees on either side and randomly placed clapboard houses on parched-looking lawns. I paused to change my carrying arm. At that moment, a battered half-ton truck careened across the empty road onto the wrong side of the street where I was standing. It slid to a halt on the dusty verge right beside me. The brightness of the day made it impossible to see inside the cab.

"Patrick?" A voice rang out, friendly, amused, from the dark interior. It wasn't really a question. Who else could I be? Before I could answer, a short, barrel-chested man leapt from the truck. He was in a tattered western-style shirt with pearl buttons holding down breast pockets above ancient, faded blue jeans. A pair of manure-caked, suede cowboy boots—rough-outs I soon learned they were called— poked out from beneath the tattered cuffs of his jeans. Best of all though was the bruised and stained ten-gallon hat that completed the cowboy look, tilting back from a grinning, weather-beaten face.

"Tom?" I tried to match his welcoming tone, but I wasn't equal to the grip of the solid square hand that was now vigorously pumping mine.

"I come down to the station half a dozen times now," he explained. "As usual the train wasn't on time, and no one cared to guess when it might show. I've been at the barn, so I thought I might as well swing past again."

I climbed in the half-ton beside Tom, but my shyness made it difficult to know what to say next. I had been waiting for this moment ever since I opened a letter addressed to me three months ago in Ottawa. It had been from Tom McCready in response to enquiries made unbeknownst to me by my father, who had spent several summers with Tom while doing geological mapping in the mountains in the 50s. Tom had been his guide and outfitter in those final days of pack horses before the helicopter took over as standard transport in geological operations. With a cook, an offsider (the helping hand to pack a horse), and twenty horses to carry supplies, my father had spent many weeks camping in the wild with Tom. Most of my father's stories about Tom fell into three categories: bear stories, horse stories, and dude stories. I was awed. This man was undoubtedly the most famous packer and wrangler throughout the region. He had not only chucked the odd rock or two to drive off marauding grizzlies, but he could play the theme music of Old Rawhide on the harmonica. The story of the French dude and carnivorous frogs up the Snake-Indian had left me weak at the knees from laughing so hard. "I'll need some help around the barn, and there may be some trail work," the laconic letter had said. And now, here I was, sitting beside a legend, quite obviously a dude.

"Keep your eyes and ears open and your mouth shut," my father had counselled. "No sense making a fool of yourself before you actually do anything." Exactly, I thought. Nothing's worse than having someone ask a lot of damn-fool questions all day. So I didn't ask any of the burning questions I was longing to ask—Is there going to be trail work this summer? Will I get to ride and pack horses? "My father is doing some work in the Northwest Territories," I finally volunteered.

"Is he now?" said Tom kindly.

We drove the few blocks to Tom's house. It looked like all the others, a small bungalow set well back from the road. Unstained cedar fences divided the property from its neighbours. Tom grabbed my duffle from the back of the truck and led the way around the path to the back door, which served as the main entrance. A small back porch allowed the removal of dirty boots. Nearby stood a chicken-wire cage. Its gate was open and in the entrance squatted the most enormous rabbit I had ever seen. The beast seemed to give me a malevolent glower as I followed Tom into the house.

"This is my wife, Fay. Fay, this is Patrick." I shook hands with a small, attractive woman with flowing brown hair and what I thought were worried-looking eyes. Like Tom, she was wearing a flowered shirt and blue jeans.

"We were so concerned when the train was late. Was the trip all right? I kept reminding Tom to check at the station, but he is so forgetful. I've been keeping some pie warm and then I took it out. I think it still might be warm enough. Sit down here. Would you like some ice cream with it?"

Feeling a little overwhelmed, I sat down and had no problem accepting the big slice of lemon pie Fay put down before me. "The train ran into a car at a crossing near Moose Jaw," I explained. "I got out for a look. It was a real mess, really horrible. The car was still wedged under the front of the train. It took a long time to remove the body and then the car, and get going again."

"Oh, what a terrible thing to have seen," said Fay, and I could see she was concerned about me as well as upset. I had forgotten the whole thing in my excitement at seeing mountains for the first time, though I had never seen a dead body, or what there was of it, before. But I was young, and the horror had been short-lived. I instantly regretted making Fay so distressed with my story.

At that moment, a small blonde girl came running through the back door. "Daddy, Caesar is out again!"

"That goddam rabbit," said Tom. Then he smiled and shook his head. "Never mind, Tessie. Come and say hello to Patrick. He's going to be staying with us all summer."

Tessie, who looked about five and had the same long brown hair as her mother, was a real cutie. She clearly had both her parents wrapped around her baby finger. And probably would soon have me too, I thought.

"We better head down to the barn," said Tom. "Patrick, you can spend the night here with us and then we can fix you up a place down at the barn tomorrow. Coming Tess?" The three of us squeezed into the half-ton and drove the few blocks to the northern edge of the town. Here, the road was bordered with stables and pony barns. Beyond rose steep, pine-covered hillsides. To the east, the mountains had begun to glow pink in the lowering sun. I jumped out to open the gate.

"That's a beautiful horse," I called to Tom as I admired a mare in one of the paddocks, closely followed by a spindly legged foal. Tom chuckled gently and gave me a sidelong glance. It took me a few weeks to develop a more critical eye for the permutations and combinations that the shape of a horse could take. The mare was Bunty, possibly the fattest and most ungainly of Tom's stock of more than fifty horses. She may also have been the meanest. She certainly seemed to have it in for me, starting the next day with a sharp bite to my shoulder as I brushed past her.

Tom showed me around the various buildings he collectively called *the barn*. There was a blacksmith shed containing row upon row of horseshoes and a separate stable with several stalls and a hayloft on top. The largest building housed the gear: riding and pack saddles, lanyards, halters, and bridles. It smelled agreeably of dust and leather. An adjoining room contained modest living quarters with another loft above piled high with tents. The whole set-up made me tremendously excited. I was going to be a real cowboy! What more could a fifteen-year-old aspire to? Tom McCready was already my hero, and I was prepared to do whatever he wanted of me. Just let me get on a horse, and I would be fulfilled for life.

I didn't sleep that night. Visions of the Old West, pack horses, bucking broncs, cowboy boots, and hapless dudes filled my mind. I expected it all to start tomorrow. In the morning, I moved into the loft among the tents, and Tom took me out into the main paddock. "Patrick," he said, "I wonder if you would mind spending the next couple of days raking out all the paddocks."

Well I raked. And I raked. To break the monotony, I shovelled manure into a pile. Then I raked some more. It was hot and dusty. My arms ached, and blisters appeared on my hands. Two and half days later, Tom called me into the tack room. He had set up a barrel on four legs. "Time to learn how to pack," he said. He threw a blanket over the barrel. "Always remember to put it farther up on the withers than necessary to start with, then slide it back into place to keep the natural alignment of the horse's hair." A wooden pack saddle, the horse's name carved into one of its sides, was next. "Do the forward girth up first, and remember, most horses don't like the back girth to be too far back. It makes them buck." He unwound a small-diameter hemp cord, called the basket rope but actually two separate ropes, from the two crossbucks mounted above the side pieces. Each line of the basket rope was left hanging on either side of the barrel. With a deft flick, Tom tossed each line around the crossbucks so that they hung down in two loops that were ready to receive a pannier or pack box. He demonstrated how to take the weight of the pannier while tightening the basket rope around it. Then, bringing the loose end from under the pack, he fastened it around the basket rope with a cunning slipknot that self-tightened to hold the panniers tightly against the crossbucks. I was delighted by the cleverness and simplicity of it. Was that all there was to it? Packing a horse wasn't going to be difficult after all.

"Now comes the top pack," continued Tom as he threw a sack of bridles between the panniers. "And next the tarp. Be sure the folded edge faces forward. Now, get over on the horse's right side and act as my offsider." He unhooked a long coil of much heavier rope, the end of which was attached to a wide canvas girth. Hmmm... maybe there was more to packing a horse than I realized. "This is the lanyard rope," he said. "You use it for the diamond hitch."

The legendary diamond hitch! I had read my westerns. I could vaguely remember hearing the term used by the cowboys who had packed the horses on one of my father's early Rocky Mountain field trips when I was about seven. With the girth and most of the lanyard coiled loosely on the floor, Tom laid the rope's end along the length of the imaginary horse's back. Then picking up the girth, he tossed it over to me. "Here you go. Now swing it back to me underneath the horse. Don't let it bang the legs or the whole shebang will get bucked off." There followed a rapid, baffling sequence of moves not unlike a complicated magic trick. "The diamond is being made now," said Tom and proceeded to loop one strand of rope under another. I couldn't see any diamond. "Pull on that rope while I take up the slack. That's it. I'll do the same for this side. We finish up by fastening the end here."

With the three packs, a pannier on each side, and the top pack protected by a green tarp, the barrel was scarcely visible now. The lanyard rope wound neatly and symmetrically around the load holding everything tightly in place. On the top of the load, at its exact centre, a perfect rope diamond had been formed. Two of its points were aligned with the horse's back, and two were at ninety degrees. Somehow, the lanyard, now a complex labyrinth of rope, contained loops that each pulled equally at the four angled corners of the diamond. It was like a miracle. It was certainly a work of art. I was awe-struck by the beauty of it and particularly impressed by how the rope tightened against itself, preventing it from loosening under the stresses imposed by the horse's movements.

There was not a single knot that didn't simply pull undone. No one who works with rope ever uses knots that are difficult to undo. The whole diamond hitch ended in a single slip knot that was protected from being accidently pulled undone by a simple half-hitch. Unpacking the horse starts with untying the half-hitch followed by a single sharp tug to release the slip knot. With no further tension holding the diamond hitch tight, the lanyard is left lying loosely across the tarp. The girth can then be released and the lanyard, with no tangling, can be coiled as it is pulled away from the pack. After removing the tarp and top pack, the basket ropes have their own half-hitches and accompanying slip knots to release, enabling the panniers to be removed. Even the two pack saddle girths release the same way. There is a knack to pull the pack saddle off the horse so that the girth straps fall down over its top, making it ready for storing.

Tom ran through the procedure again, then left me to practice for the rest of the day. It was enormously satisfying to gradually perfect the rope manoeuvres by myself, each time learning another small trick that kept the ropes tighter or made the sequence of loops easier to perform. The success of a pack, Tom had told me, could be determined by the shape of the diamond. The poorer the shape, the less likely the pack would survive for long. By the end of the afternoon, my diamonds looked perfect. But the following day, when I graduated to a real horse, I discovered that packing a barrel was much easier. A barrel didn't tread on your feet, turn and bite you, or constantly change the diameter of its girth. "When the horse feels the girths tightening, they can blow themselves up," explained Tom, as he surveyed all three girths hanging loosely below the horse's belly.

"When you're finished packing they relax, and the girths become too loose to hold the pack." He showed me how to recognize when the horse was attempting to pull off this blowing up trick and how to give it a judicious jab with your knee while tightening the girths. The jab causes the horse to exhale and the girth gets tightened before it can recover.

Tom gave me a lot of different jobs to do in those first few weeks. One day, I was waterproofing canvas tents up in the loft when I heard a wild whinnying. Looking out the casement window, I saw a cowboy unsaddling a superb, but skittish, chestnut. The two were perfectly matched. The cowboy was lean and hardened-looking in a black hat, worn leather chaps, boots and spurs, the complete western look. The horse was over-excited and lathered in sweat but had the aristocratic air of a racehorse. It was a potentially dangerous moment, but the cowboy deftly avoided the prancing hooves as he expertly got the saddle off. Soon I heard the newcomer moving down below. Unsure what to do, I continued brushing canvas until I heard him coming up the stairs. "Hello?" I shouted, not wanting to be taken for an intruder. Close to, I saw a man in his early twenties. He had a saturnine complexion and a thin mouth that gave him a cruel look, but he introduced himself cheerfully enough. He was Glen, and he was going to be working the rest of the summer for Tom. He had arrived on Tom's most prized stallion, Abu Abdu Allah, or Abby for short.

It would be charitable to credit my relationship with Glen as the reason I learned the cowboy art as thoroughly as I did. His personality turned out to fit his appearance. I soon learned that most of what Glen told me was either a deliberate lie or at best a half-truth. And always malicious. He would watch with interest as I applied the last half-hitch on a fully packed horse, then inform me that I had put Becky's pack saddle onto Goldie, something he had known from the outset. Glen was a veteran at Tom's and knew all the horses, but it would take me a while to recognize them all by name. I'd grasped right away how important it was to match the name on the pack saddle with the right horse. An ill-fitting saddle could result in serious sores, but I wasn't about to invite Glen's derision by asking him if I had selected Becky correctly from a full corral of horses before I started.

After my cowboy education was a little more complete, I made the most of an opportunity for revenge. We had been packing steadily in and out of Tonquin Valley for some weeks. Glen and I ran one pack train of twenty or so horses, while Dave, a full-blooded Cree, worked another with Tom. One train went in full while the other returned empty. It was a day trip each way. In the Tonquin, we had a lean-to set up beside a small creek (always pronounced *crik* if you wanted to sound like a cowboy). The day started at 4:00 a.m. Before doing anything else, the horses had to be found, unhobbled, and driven into a makeshift corral. It was always cold. When I worked with Tom, we did this job together. Two people made it far easier. Glen, however, had other ideas. He liked to send me out alone while he prepared the packs. After one particularly long day, we turned in only to be woken by the lean-to collapsing on top of us. It had started snowing, big heavy wet flakes. Tired and cold, we struggled half-naked in the soupy slush, trying to get enough snow off the roof to set it back up. When all was right again, we had barely started to warm up inside the down bedrolls when the alarm announced the beginning of a new day.

"You go ahead, Patrick, and get the horses. I'll get the packs ready," Glen told me. On this particular morning, the order seemed blatantly unreasonable. We lay in the dark arguing.

"No way," I told him. "It's your turn. I'll get the packs ready."

"You don't know how to make a proper pack, fer Chrissakes," said Glen. "Remember when you tried to load the kettles on Ollie?"

He was reminding me of an incident I would sooner have forgotten. The jibe was below the belt, nothing to do with the issue at hand. But at four in the morning, it was difficult to produce a balanced and rational counter-argument. In the end, highly pissed off, I crawled out of the lean-to. My last words to him were, "You bloody well better have those packs ready when I get back."

My boots were already soaked as I made my way through the neck-high salix. Dawn was breaking, and I could just make out the forms of motionless horses half a mile down the valley. When I reached them, the nearest horse was Skipper. I had grown to like Skipper a lot. He was small, but fast, and seemed always ready to do a good job. I approached him slowly and carefully, a bridle held with one hand behind my back. "Come on, sweetheart," I coaxed, "easy does it." By now, I had learned a few tricks. There is a kind of intuition in getting the right angle of approach to a wary horse, adjusting it with each slight change in the horse's movements. This morning, Skipper was having none of it. Lunging forward on his hobbled front legs, he was too quick for me. "Sonofabitch Skipper," I pleaded under my breath as I started another slow approach. This time he let me touch his neck before making another crippled getaway to the other side of the creek. I was too cold by now to feel the icy water that filled my boots as I waded after him. Skipper required three more attempts before tiring of the sport, but at last I got the reins around his neck and bridled him. With frozen fingers, I struggled with the equally frozen hobbles. Then to ensure that my soaking was complete, I climbed up onto his bare, snowy back and began the roundup.

"One, two, three, four...." I started the count as I circled around the herd. Damnation! How many did we bring in yesterday? I couldn't remember. I counted twenty-one. That felt right, but then, looking down to the end of the valley, I could just make out two more. So down we went, Skipper and I, getting nearly all the way there before it became clear that my two renegades were in fact a pair of peacefully grazing moose. Dawn had thoroughly broken by now. My jeans were soaked through from the still snowy brush, but Skipper's flanks were beginning to steam, and it was now warmer on him than off. Back at the herd, I removed the hobbles and began driving the horses back toward the makeshift corral near the lean-to. A glance showed me that no pack preparation had yet taken place. With the horses safely in the corral, I remained on Skipper and circled round to the front of the lean-to. Perhaps I was almost hoping for the sight that confronted me. There, as snug as a bug in a rug, lay Glen, completely invisible inside the luxuriantly thick bedroll. Steam rose from one end as the whole package moved gently and rhythmically up and down.

In a cold fury, I drove Skipper forward into the open lean-to entrance. Glen awoke to the pounding of hooves inches from his head. Wide-eyed and open-mouthed, he tried to spring up. But I had already slithered down from Skipper's back and was pulling his naked form out of its warm nest. I suppose it was cowardly of me. I held all the aces. In normal circumstances, Glen could probably have thrashed me with one arm tied behind his back. Instead, confused and bewildered, startled out of a deep sleep, he found himself being rolled about stark naked in muddy, icy slush. I showed no mercy. Every time he tried to get to his feet, I pushed him back down and for good measure pelted him with another barrage of wet, filthy snow. Skipper seemed to enjoy the spectacle too. He stomped and whinnied with abandon, adding a whole new dimension to the sport and confusion. Finally, when I felt enough was enough, I went off to start making up the packs. I didn't look behind me, and I never had any further trouble with Glen. I refrained from telling Tom about the incident, and I'm sure Glen's pride stopped him from saying anything. But Tom must have sensed the hostility. For the remainder of the season, I found myself packing with Tom. I also watched the

Magic Travels

relationship between Tom and Glen deteriorate. It wasn't long before Glen was fired.

∞

Tom's right-hand man was Bill McKinnon. A movie director could have done a lot worse than casting Bill as the lead in a western. Not in one those melodramatic John Wayne horse operas but in an honest, realistic portrayal of the Old West. Bill was the epitome of a true cowboy, tall and lean with a bronzed face. You could see the whole horizon between his legs. His right hand was missing its third and fourth fingers, the result of an argument with a bulldozer in his catskinning days. Bill liked to dramatize his innumerable stories with a lot of vigorous gesticulation, and this hand would wave about in a permanent bullshit gesture. But who could know the truthfulness of Bill's yarns? They were always funny and always accompanied by his unique honking laugh whenever he paused to draw breath. Bill was also generous to a fault. Time and time again, I watched him pass money to drunken ne'er-do-wells, white or Indian after being accosted in the street. He obviously had a reputation as an easy mark.

Whether Bill ever got any of his loans back I never heard. I was walking down Geikie Street one day, clad in boots and hat myself by now, when a voice called, "Hey cowboy, could you loan me five bucks? Just for a short time. Really." The speaker was an unshaven, weaselly-looking derelict. I felt inordinately flattered to be addressed as *cowboy*. I also had five bucks in my wallet that I didn't really wish to be parted from, but such was my respect for Bill, who would have handed at least some money over unhesitatingly, that I stopped, complied, and kissed my five dollars goodbye. A few weeks later, the same man approached me. He was clean-shaven and wearing new clothes. "Hey, you loaned me five bucks. Let me pay you back." The man shoved a note into my hand, jumped into a truck that was still running beside the curb, and drove off. Astonished, I unfolded a ten-dollar bill.

I doubt that either Bill or Tom had any grand design for my metamorphosis from dude to cowboy. Both simply gave me whatever jobs came up. Before I had even sat on a horse, Tom informed me it was time to get some boots and a hat. At least I could look the part even if there hadn't yet been an opportunity to act it. All the same, I could only laugh uneasily at the story of the pony barn operator who always asked his tourist customers if they wanted a horn with the saddle. "Do you think there's really enough traffic about?" was one alleged reply.

One evening, Bill and I were cooking supper in the barn when Tom burst in. "The horses are grazing in the middle of the airfield," he shouted. "Quick, saddle up, let's get going!" There was a small grass landing strip outside Jasper. If it were proved that the horses belonged to Tom, a heavy fine was possible. Both men seemed to have forgotten that I hadn't actually ridden before, at least not since I was seven. There were three horses in the paddock. "Take Smoothy," shouted Tom, pointing to a neat little mare. No time was lost saddling them, and soon I found myself cantering after the others down a dusty trail. This was high excitement. My adrenalin flowed. I had no idea where we were or what I might have to do once we got there. For the moment, my principal concern was to stay on my horse. The others were soon far ahead of me. When the trail opened out onto the airfield, I could see them already heading toward a small stand of cottonwoods down at the far end. Horses stood among the trees grazing on the lush spring grass. By the time I arrived, Tom and Bill had already completed the roundup and were driving the discontented animals into the range country among the terraces to the west.

This herd had been wild on the range since the previous fall. They had no intention of being put to work again. The chase was at full gallop. I could feel my mount surge into top gear. I felt the same way and leaned forward on the stirrups, oblivious of anything but my desire not to be left behind. I revelled in the speed and marvelled at the muscular power of Smoothy thundering beneath me. We were travelling over a network of minor trails that wound through lightly treed, gently rolling terrain. Any attempt to actually steer my steed seemed useless. Instead, I hung on and concentrated on ducking the low branches that constantly threatened to knock me flying. I now fully appreciated the practicality of chaps as the brush tore past my legs and the sides of trees occasionally banged painfully into my knees; chaps weren't just for the looks.

All seemed to be going well until an elk skeleton lay right in front of me, directly in the path. I would observe this phenomenon many times in the future. Horses can be spooked at the sight of death. Still flying at a roaring gallop, Smoothy took a sideways lunge to avoid the fearful object. I remained in the previous flight path. CRASH! A second later, I lay flat on my back among the bones, looking up into a darkening sky. Slowly, my breath returned. I wondered if I had the courage to assess the damage and decided against it for the time being anyway. All was quiet, save for the gentle rustle of leaves in the wind. I imagined Smoothy miles away now, continuing the chase probably not even aware that the dude was no longer on top of her. Finally, I sat up, stood up, threw up, and sat down again. No lasting damage was apparent. To my amazement and relief, I spied Smoothy, who for reasons I could not fathom, was only yards away calmly munching on grass as if waiting for me to let her take me home again. Painfully, I climbed onto her back and she carried me, gently I thought, toward the welcoming lights of Jasper. I was greeted with much merriment from Tom and Bill. They both treated the whole episode with considerable levity, but I thought I could detect a small look of relief in Tom's eyes, if only because Fay, who had joined them on their return to the barn, could now stop worrying.

My cowboy education was definitely picking up speed. Tom instructed Bill and me to round up a herd of horses that had wintered far up the Snake-Indian River. A locked and gated fire road led off from Moberly Flats where the Snake-Indian joined the Athabasca. Bill had the key, and we followed the winding, dirt road some twenty-five miles up the river valley to a warden's cabin known as Seldom Inn. Leaving the pickup there, we tramped into the bush until we found some recent tracks and soon sighted the herd. Bill made catching a horse look easy. He knew which ones to go for. Once we had two bridled, the next step was to ride bareback (*skin-ass* if you are a cowboy) and drive the rest back to the corral at Seldom Inn. "Grab a hunk of mane in one hand," said Bill, "and don't let go." I swung up, sat still for a second, and slithered down the other side. Still hanging onto the poor beast's mane, I struggled up again, grasping at anything to stay on. Meanwhile, the horse, not taking kindly to such abuse, trotted after the others, who were being expertly driven at a fast clip by Bill. He looked like he was growing out of his horse. I was being bounced unmercifully. No sooner would I achieve a sitting position when down the side I went again. The only reason I made it back to the corral was by following Bill's advice to the letter. I never let go. Nor had I spent much time actually on the horse's back. But necessity is a fine teacher, and over the next few days I learned the knack of sitting back, floppy and relaxed, letting the looseness of my body absorb the jolts. In this way, it was possible to ride at a trot from sun-up to sundown, even skin-ass if need be. It also seemed super-cool, and whether I looked the part or not, I now felt like a real cowboy.

"Okay," said Bill, "I'll be off now. You drive the horses back to Moberly Flats while I take the truck. I've got some things to do in town. I'll be waiting for you." He was gone before I had time to protest. I watched the dust from the truck settle as he disappeared around the bend. There was a stillness to the air, a closeness. It had been getting steadily hotter. I became aware that my shirt was soaked in sweat. "Well, here goes nothing," I thought. Taking my bridle and holding it behind my back, I quickly caught a pretty little mare, an easy job in a packed corral. Then, letting the reins fall, I returned to the railings for saddle and blanket. Rule number one: never let your horse go free once it is bridled. It is impossible to catch again especially while carrying a saddle in both arms. It's not even easy work with two free hands. After putting the saddle on the ground, I spent a merry half-hour floundering through the crowded corral trying to regain my mount. Every time I was close to success, she would dart into the melee of an increasingly agitated herd. Finally, I managed to catch the bridle in a lucky lunge, led her out of the corral, and tied her firmly to a hitching rail beside the warden's cabin. Rule number two: never tie a horse up by its reins. As I bent down beside her to make an adjustment to the saddle, the sky fell onto the top of my head. When I came to, I found myself stretched out on the ground with the hitching rail lying across my body. The mare had reared back and pulled the rail free from its ancient rusty fastenings. It had hit me square on the back of the neck. Even worse, the jerk had torn both reins off the bit, and she was free once more. I was a receptive listener when Tom later explained to me, "if you have to tie a horse by its reins, never use a knot. A few loose loops will do. The horse will think it's tied, and even if it pulls free, at least the reins won't be broken."

I spied another bridle hanging from a nail under the eaves of the cabin. Hoping Tom was on good terms with the warden, I took it and began wearily chasing my mare for the third time. She must have taken pity on me. After a few token getaways, she finally stood still and let me install the new bridle. But it hadn't been out of pity. The heat had become oppressive, almost ominous. The slightest exertion was an effort for both me and the horses. I opened the corral gate and rode in to drive them out. Morosely, we headed down the road at last. In such heat, the dust kicked up by the horses rose high into the less dense air, obscuring vision and caking the nostrils. I shouted; I cajoled; I threw rocks. Our pace remained like the flow of corn syrup. By late afternoon, I was seriously worried. We still had at least twenty miles to cover by my reckoning. Every step was slow and leaden, but despite this leisurely pace, sweat flowed copiously down the dusty flanks and necks of the animals. By driving my mare into the rump of the horse ahead of me and lashing out at it with the free ends of my reins, I could occasionally elicit a bedraggled trot. I felt my mount start to slacken with weariness and changed to another horse. The herd showed not the slightest inclination to disperse while I did this. They stood dejectedly in the road apparently satisfied just to have stopped for a few moments.

The unnatural heat produced an uneasiness in the atmosphere. As dusk fell and the clamminess got worse, the uneasy feeling intensified. On the skyline, silhouetted by darkening mountains, huge mounds of cumulus began to form. They towered over the peaks in an awesome display of pent-up power. By now, a kind of telepathy had developed between me and the herd. We were sharing a primordial, instinctive fear of the elements. The tension in the air increased as massive billows of blackness began to fill the sky. The last rays of the sun that had illuminated the tops of the eastern peaks were blotted out. A stab of lightning cut across the heavens. Seconds later, a maelstrom of thunderous sound engulfed us. Starting with the sharpness of a cannon shot, it continued in a barrage of noise that shook me to the gut, tailing finally into a series of menacing rumbles.

Panic shot through the herd. I could feel it sweep down the line in front of me until it hit my own horse with a surge of raw energy. In one accord, my charges broke into a wild gallop. At the same moment, heavy rain began to pour down in a torrent. The blackness of hell seemed to have fallen, broken only by jagged thunder bolts cracking through the dark, illuminating the surroundings in staccato flashes. In seconds, the road became a river down which we hurtled amid a confusion of noise and driving rain. Fierce squalls whipped overhead branches into a frenzy. The cracking of falling trees intermingled with the crash of thunder. I no longer had any control of the herd. I could barely even see the horse in front of me. Blindly, I hung on, giving my mount full rein for fear of losing the lot of them.

It was impossible to do anything else. I wondered where the corral was and how on earth I was going to drive the horses in when we found it. Had we already passed Moberly Flats? There was no way of knowing. For what seemed like hours, the mad ride streaked through the darkness. The initial panic had resolved into a fast, steady gallop, with the herd holding together in a kind of earnest determination to reach... where exactly I could no longer tell. I comforted myself with the notion that they knew the way, while I concentrated my efforts on peering ahead into the emptiness. Not that there was anything to see. The rain streamed down my glasses, giving me all the visibility of swimming underwater.

Without warning, the pace altered. My horse stumbled off the road, nearly leaving me behind in the mud. I became aware of curses, barely audible above the shriek of the wind. To my heartfelt relief, I saw a banshee-like figure in a streaming slicker wildly waving a storm lantern. "WHOOOOOOOOOAH! Get in there! Goddam it! C'mon! In you go!" It was Bill. He had been true to his word and come back to meet me. Turning sharply, the horses thundered into the corral, and Bill pulled the poles across the entrance. I slid the saddle off my horse and fled away from the stamping herd. Grabbing my arm, Bill led me through the blackness into a small log cabin. He held up the lantern and inspected my sodden, bedraggled form. "You know," he said, "I kinda figured you might be having a little trouble."

∞

I finally learned what the summer really had in store for me. Tom had a major contract with the Alpine Club of Canada to supply them with all the tents, food, and equipment required for a huge climbing camp to be set up in Tonquin Valley. This was an enormous undertaking. We started by setting up a base camp at View Point beside the road at the foot of Mount Edith Cavell. Everyone called the place View Point even though it was only one of several thousands of other viewpoints signposted on the National Parks' highways. This particular viewpoint was a small pull-over with a magnificent vista looking up the Astoria River Valley. We built a corral and set up camp among the trees. Nearby, a brook flowed from a small spring, giving the sweetest water I have ever tasted.

The work was divided into three phases. Two weeks were spent packing in all the gear. Then the alpiners were to arrive, hundreds of them. Their climbs in the surrounding mountains would last another two weeks, and we would have to keep up a constant supply of food. A final two weeks were needed to dismantle and pack all the gear out again. Two full pack trains would be required every day of the six weeks, one going in while the other returned. Needless to say the days had to start early, usually well before dawn, and long hours would be spent in the saddle.

To add to the challenge, it was one of the wettest summers Jasper had ever known. Each morning, the sun rose into a cloudless sky, and at 11:00 a.m. sharp, or so it seemed, the sky would cloud over and dump torrents on us for the rest of the day. It was impossible to stay dry. If the rains didn't get you, crossing the swollen rivers did. The horses hated plowing their way through the swirling waters, slipping, slithering, barely able to keep their footing in a stream bed that was probably a shifting mass of mobile boulders. We had to be lucky to get a whole train across without one mishap or another

The trick was to keep the horse pointing up-current at about forty-five degrees, get your legs up around your ears, and hang onto the saddle horn for dear life. If the current caught your mount broadside, it could easily lose its footing altogether and along with its rider be swept downstream. Occasionally, a horse would swim for it, ensuring a total soaking in glacial waters. One such time I had no choice but to abandon ship as my horse swept past a grassy bank. The result was a miserable walk in soaking gear, complete with chaps and spurs, only to find the animal grazing peacefully downstream.

But for a teenager who wanted to be a cowboy, this was all adventure; this was bliss.

The next step in my education came a few days later. Tom had grazing rights between his barn and the foot of Pyramid Mountain. "Just saddle Abby and drive them lot to the range," Tom instructed me one day. This was an honour. Abby was a beautiful, but high-strung animal, and I had been eager for a chance to ride him. Unfortunately, the herd contained Bunty, the only horse that seemed to want to do me serious harm. She never lost the opportunity to take a sharp bite or give a well-aimed kick whenever I walked past her. She was a fat, unlovely beast with a personality to match.

Oh, how I felt like a real cowboy as I swung onto Abby's back and started driving the herd down the trail. True to form, Bunty wasn't in a hurry. She deliberately placed her backside directly in front of Abby's nose and ambled forward as slowly as possible. He quickly grew difficult to control, chomping on the bit and skittering around her rump. Bunty responded by insistently waving her tail in his face and, well, uninhibitedly exposing herself. Try as I might, I found it impossible to drive her farther into the herd, which was now a considerable distance ahead of us. This was worrying. A road crossing was coming up, and the route was often crowded with tourist traffic heading to and from Pyramid Lake. I needed to be with the herd.

Luckily, they had all crossed safely by the time I arrived, still hurling abuse at the placid, tail-wagging Bunty. Abby, by now, had taken to snorting and was prancing about insanely. Cars stopped politely on either side of us. So did Bunty – right in the middle of the road. Abby, still cavorting like a cat on a hot tin roof, decided to take charge. Ordinarily, I looked forward to this crossing. I liked to imagine the tourists admiring a real live cowboy "jus' doin' his work." What happened next was totally ruinous to this fatuous self-image. With a kind of strangled moan, Abby forced the reins through my hands in a desperate, frantic lunge. Before you could say "Bob's your uncle," I found myself sitting on top of not one but two horses. Or, more accurately, I was on top of Abby, and Abby was on top of Bunty. Surveying the scene from my vantage point, I could see the pale, gasping face of a middle-aged woman in the front seat of the closest car, her young wide-eyed daughter beside her. I do believe to this day that I saw that woman faint.

Confused and red with shame, I clambered down from the lustful pair. I dodged and ducked to stay clear of the excited beasts, but there was little else I could do. Abby was clearly not going to be thwarted from his objective. From the ground, I could fully appreciate the woman's shock. The full extent of Abby's manhood in furious action, accompanied by the clamour of randy whinnying and the desperate clatter of hooves struggling to maintain a precarious balance on the smooth pavement, was a stupendous spectacle.

The line of cars on either side had grown. Mothers were covering the eyes of their little ones. Fathers were getting out for a closer view. Trying to look as though I was in control of the situation, I shouted curses and frantically tried to beat Abby down from his mount. Poor Abby! The result was unwanted and forceful coitus interruptus. A fireman's high-pressure hose couldn't have sprayed the area as quickly. The windshield of the nearest car received a good portion, but I was not about to hang around and apologize. Bunty, presumably in frustration, had galloped off down the trail. Abby was still prancing about in uncontrollable torment. With some further damage to my dignity, I managed to scramble back on the saddle and made my clumsy getaway, thundering into the safety of the bush after the rest of the herd.

The job completed, I rode back slowly, thoroughly depressed. What on earth was I going to tell Tom? The incident would surely be seen as a demonstration of gross incompetence. To make things worse, I had an idea that Tom kept just on the edge of the right side of the law with his grazing rights. Now, he would have me to thank for drawing attention to his operation by supervising an obscene display on a prime Jasper tourist route. What about Bunty? Could she have received even a tiny bit of Abby's bounty? Would Tom want her to be pregnant? As for Abby, would the Battle of Pyramid Lake Road make him even more unmanageable? And me? I was already the butt of enough jokes. Why, just the other day Fay had watched me run for my life from a rabbit.

∞

Tom had asked me to paint his backyard fence. Carrying an open gallon of red stain, I stepped off the back porch unaware that Caesar, Tessie's grotesquely large pet, lay in wait for me beneath the steps, ears flat along his back, hackles up, his cruel eyes narrowed into mean little slits. Just as I crossed the walk onto the lawn, he sprang, landing on all fours on the back of my leg. There, he held himself in place with a mouthful of my jeans while lashing out with his taloned hind legs in a series of violent kicks.

There are people who don't believe this story, but every word is true. If you have never been attacked by a man-eating rabbit, don't sneer. It could happen to you. In later life, I have stood my ground in front of a bull moose and faced down venomous snakes in Australia. I have even contended with the wrath of a five-hundred-pound bearded seal who resented me coming through his breathing hole to dive under the Arctic ice cover. I have fended off polar bears and grizzlies with well-aimed rocks. One expects to be attacked by a bear. It happens. But a rabbit? That is terrifying! It wouldn't have been so humiliating if I hadn't given a shrill yelp of terror as I leapt panic-stricken a good six feet, or so it seemed, into the air, carrying the furious rabbit with me. I swiped at Caesar with the can, dislodging him but spilling red stain over both of us. To make my disgrace complete, I made a mad dash with the now bright red rabbit hot at my heels. He cornered me against the fence. I could see his mad, rolling eyes searching for my throat as he gathered his body together for a renewed spring.

Fay saved my life. From the kitchen window, she had seen Caesar attack and had rushed after us. Before the beast could deliver its death blow, Fay pounced. Grabbing it from behind by the scruff of the neck, she hurled the struggling creature to the other side of the fence. If only she hadn't been laughing quite so hard.

As the summer drew on, there were no signs of Bunty becoming pregnant, and I learned to accept my reputation as the cowboy who was chased by a rabbit. Years later, I visited Tom and Fay and, with mixed feelings, learned that the rabbit story was now thoroughly ensconced in Jasper lore. But I never did tell of the commotion that had occurred on Pyramid Lake Road that one hot day in July of 1963.

∞

Shortly after we started packing into the Tonquin, Tom realized he needed another hand. I went with him on a long drive to an Indian Reserve outside Hinton. Broken cars lay among grotty shacks. Tom pulled over in front of one of these tumbledown homes.

"Dave?" he shouted. An elderly brown face appeared at the door. It creased into smiles on seeing Tom. "You want some work?"

"Mebbe." The reply came out slowly, tinged, I thought, with an infinite sadness. Dave came out, climbed into the truck, and started to roll a cigarette.

"You want to bring some gear with you, Dave?" asked Tom. "We've got several weeks of work."

"Mebbe." Dave got out and disappeared into his shack. Half a minute later, he re-emerged with a grubby pillowcase half full of belongings.

I really liked Dave. He was the first person out there I dared to try some magic on. We were alone together in the cook tent at View Point. Dave's newfound wealth had enabled him to change to tailor-mades. "Let me show you a trick with one of those, Dave," I suggested. Before he had time to say, "Mebbe," I snatched his lit cigarette, disappeared it, produced it again from my knee, then pushed the red-hot tip into my fist. I opened my hand. The cigarette had vanished. Dave paled in astonishment, then slowly his mouth cracked into a huge grin, and his head seemed to sink into his shoulders until his ears were all but buried. To my alarm, he began to shake. No noise, no chuckling, no guffaws. His shoulders pulsed uncontrollably up and down on each side of his head. Dave was laughing.

After that, every time Dave pulled out his cigarettes, he would gravely hand them to me for a repeat performance. He never seemed to tire of the effects and never showed the slightest inclination to know how they were done. In return, he taught me the Cree alphabet. This syllabic writing system, invented by a missionary (James Evans) in the 1830s, consisted of nine glyph shapes, each of which could be displayed in four orientations that signified the four fundamental vowel sounds – ah, eh, ee, oh. Each of the nine shapes represented a consonant, with the following vowel indicated by the orientation, for example, pa, peh, pee, poh, or ka, keh, kee, koh. In this way, the Cree language could fairly easily be written phonetically. Evans taught the Cree to write by demonstrating with soot on birch bark and became known as the "man who made birch bark talk." Dave wrote it all out for me on a page in my diary. Evidently, the script had been modified over the years as Dave's version had fourteen separate glyphs and a few undefined ticks. Studying the page while writing this account, I spied a group of four glyphs. Working them out from the Cree alphabet, I realized he had shown me how to spell my name – Pa T Ri K.

It was an accepted fact at that time that no Indian working for a white man was ever reliable for long. Over and over, I heard the same refrains: "They just don't understand that a job's gotta start on time. You never know if they're goin' to show or not." To my dismay, Dave wasn't an exception. Every time Tom paid him, he disappeared. On one of these occasions, during the Jasper rodeo, I found Dave behind the stands, drunk and decidedly the worse for wear. Blood was streaming down one side of his mouth, and several teeth were absent from their usual places. He was surrounded by women in much the same condition as well as a multitude of dirty children. Grabbing my arm with enthusiasm, he handed me his cigarettes. I did a show for them all amid whoops of drunken laughter.

It was one of Dave's no-show days when I first found myself in sole charge of a complete pack train. The alpiners had all arrived and gone in on foot to the Tonquin, which now looked like Tent City. The weather was not being kind to them. The Tonquin seemed to trap low fog, which then produced a steady drizzle. Many of the tents now wallowed in a sea of mud. The horses were getting rundown and showing signs of distemper. To us, the alpiners were dudes, of course, and crazy ones at that. The idea of climbing mountains for sport was not an activity that cowboys could identify with. Bill had a low regard for the Tonquin. "You know," he told me as he threw a well-aimed rock at a fully loaded horse struggling to get out of a bad mud hole, "you know, if I was a tourist, you know... just a plain tourist with lots of money... and I paid someone good money to show me the Rockies... and if someone I'd actually paid money to...

you know... to take me to the prettiest place in the mountains... and if I was brought out to here... I reckon I'd shoot the fellow."

On the morning in question, a load of food was ready for packing as well as two huge, perfectly smooth cooking kettles. Several smaller pots were nested inside of each. Exactly how to pack them was a puzzle. There was little for the basket rope to hold on to. Even wrapped in tarps, the kettles made a difficult, awkwardly shaped pack. I chose Ollie to carry them. He was a huge animal, extremely thin, with an exceptionally long neck and high withers. It was generally believed that he was a cross between a moose and a giraffe. Whatever the case, he had a tiny brain. But he was a powerful animal and, having little imagination, made quite a reliable pack horse. He was even a reasonable mount if you didn't mind his looks and were able to get a bridle on him. Ollie had only to extend that long neck of his to make his head impossible to reach without standing on a pack box. Down at the barn, the trick was to lead him into the low-ceilinged stable first. Ollie would see the bridle coming, jerk his head up, and nearly knock himself out. The bridle could then be slipped over his head on the rebound. It was an obvious trick as well as a shabby one, but he never learned.

Because Ollie was so tall, he made a useful caboose at the end of the pack train. The extra height gave his rider a good view of the packs, so problems could be spotted and repairs made before a disaster happened. However, with most other horses you could more or less relax, even snooze or read a book, and trust your mount to follow the others. Not so with Ollie. You had to steer him around every bend in the trail. Otherwise, he would just plow on in a straight line until a tree halted further progress.

As our strongest horse, Ollie usually got saddled with the heaviest loads. One time, we packed him with two weighty panniers of canned Carnation milk. I led him fully loaded into the corral to wait with the others while we finished the packing. Ollie ambled toward two trees growing side by side in the centre of the corral. Oblivious of his load and without any malice, he decided to walk between them, but the gap was far too narrow. Brought to a halt, Ollie, unperturbed, gathered himself together and gave an enormous push. The force was enough to break through the panniers and crush the cans. We watched in awe as magnificent white arcs of precious condensed milk fountained across the corral.

After a good deal of improvisation, I had managed to get the kettles onto Ollie, but this was not the end of it. Dave had been advertising my magic skills, and the alpine community had requested a performance. I threw my magic case on as a top pack, and we set off at last. I could see that Ollie didn't think much of the kettles. They slapped against his sides as we walked, making loud clanking noises that perplexed him. He kept swinging his head around and eyeing them suspiciously. Nevertheless, we progressed down the trail without mishap until Popeye, who always insisted on taking the lead, decided it was his time to stop and, as always, admire the view. I was used to this and knew that in time he would tire of gazing at the scenery and continue on the way. In any case, the view was marvellous, the sun was still shining, and my heart was soaring with pride in the responsibility I had been given to single-handedly look after an entire pack train. I could wait.

At this point, the narrow trail did little more than provide a temporary interruption to the steep slope of the cliff. I marvelled for the umpteenth time at how sure-footed these mountain horses were. "You'd never dare bring one of them pony barn horses out here," Bill had told me. I could believe it. Several times already, I had felt secure on a mountain horse in places I wouldn't have dared to navigate on foot. No, I couldn't grudge Popeye his peaceful interlude. But I didn't want him to get complacent, so just to show him who was in charge here, I threw a token rock and shouted a few more curses, then settled back to admire the view and wait out Popeye. A droning noise interrupted my reverie, and I turned my attention to the sky, expecting to see a plane. Instead, I saw a cloud of bees. And they were angry. My last rock, which had gone wide of its mark, must have landed on their nest.

"Popeye!" I screamed as the buzzing grew louder, angrier, and closer. My new tone got through to him. He started off, but the bees had already picked their victim. Tall, eye-catching Ollie. I watched helplessly as the swarm gathered and in a single mass descended on him like a blanket. At the first sting, Ollie shot straight up with his back arched like the proverbial Halloween cat. He landed on his front legs, kicking wildly and doing a remarkable imitation of a handstand. The kettles didn't have a chance. They flew off together in a graceful arc, the magic case between them, leaving behind a mess of tarps and ropes. Miraculously, the case landed flat side down on a scrub pine and slid gently to the crevice between the tree and the cliff. The kettles collided with the mountainside and began to roll, gathering speed and momentum as they plowed through the light bush. Soon they were out of sight, though I continued to hear them, the sounds dying away until only an occasional clink marked their progress to the bottom of the canyon.

In the meantime, Ollie and all the horses in front of him had taken off at full speed down the trail. I could do nothing except watch Ollie's frantically bucking form disappear around the next bend, the bees still swarming him. I rescued the magic case and tied it temporarily on the back of my saddle. There was no time to retrieve the kettles, now many hundreds of feet below. They would have to wait. I shouted at the remaining horses, and we set off. On the narrow trail and with full packs, I didn't dare push us faster than a walk. Eventually, the valley rose to meet the trail. The ground was flat once more, and there stood Popeye, Ollie, and the others, peacefully grazing. Streamers of rope hung from Ollie's pack saddle, many of them wrapped around his legs. He could only take little mincing steps but didn't seem any the worse for wear as I tidied up the lanyard and basket ropes and resecured his saddle. Next, I selected Goldie, a peace-loving horse who would not be easily spooked. I unpacked her completely, added my precious case to her load, packed her up again, and off we went once more.

Near noon, with still a long way to go, it began to rain. I reached around to untie my slicker from the back of the saddle. It wasn't there, though the lanyards were still tied. It must have slipped out after I removed my magic case to pack it on Goldie. By now, it was probably lying on the trail miles behind us. My clothing was quickly saturated, and the afternoon grew progressively colder. The excitement I had felt at the start of the journey was evaporating. The glamour of the job could not survive the freezing mud that splashed across my face, kicked up by the horse in front of me. For hours, I sat hunched against the rain, shivering uncontrollably. Even my hat began to leak, and the damp made it stretch and sit heavily on my ears. The horses plodded their weary way through the quagmire that was supposed to serve as a trail. I knew we were approaching an especially bad section of boulders and mud, a combination that always resulted in much slithering and sliding. Some repacking was often necessary after this stretch. I prayed we would make it through without that. I doubted my ability to repack a horse by myself in the pouring rain with fingers so cold they were barely able to hold the reins.

Just as we were taking the final steps onto firmer ground, there in the middle of an especially large mud hole I spied a single horse blanket. No matter how tight and snug you make a pack, wet rope can stretch enough for a blanket to work its way slowly out from under the saddle. The pack still looks perfect, but now there is no protection for the horse's back. A complete repacking from start to finish is essential. Groaning, I made a grab for the blanket from my saddle and pulled the muddy, sodden mass on to my lap. My own horse was floundering nearly up to its belly in the mud, so at least it wasn't far to reach. Urging my mount out of the mud, I trotted my way through the sodden animals and brought the train to a halt. Wrapping (not tying) the reins around a branch, I walked the length of the line. With panniers and tarp, it was impossible to see the absence of a blanket without checking underneath each load. The blanket belonged to Goldie. Maddeningly, the pack itself was in excellent shape, the diamond still in perfect form. That didn't matter. Repacking from start to finish was unavoidable.

As I worked at the first half-hitch, I became aware of the smell of wood smoke. I knew there was a warden's cabin around the next bend, but until now it had always stood locked and empty. Comfort and help were near. I stopped unpacking Goldie and drove the train forward. We rounded into a clearing containing a most welcome sight. There was the cabin, a spiral of smoke rising from its chimney. A couple of slickers hung on pegs under the eaves, and a pair of horses was peacefully munching on oats in the corral. The rain was now heavily mixed with snow, making the rustic scene all the more inviting.

"Just got a pot of coffee on the stove. You work for McCready? C'mon in out of the wet. Christ, it's cold. Shove the horses in the corral, they'll be okay for a while." I had never met the warden, but he was already on the porch, having heard my approach.

"I've got to repack a horse," I called back.

"Sure, sure, but there's time for that. I'll help you after. Come in and get warm first."

I didn't need much persuading. A moment later, I was standing beside a pot-bellied stove humming from the roaring fire inside it. Sparks of red glowed from the cracks. A saucepan of coffee was boiling merrily, cowboy-style. A huge German Shepherd welcomed me as though I was his last friend on earth. Steam rose from my clothes, fogging the one small window that looked over the corral.

"Why don't you put those on?" The warden indicated a pile of dry clothes. I could see a warm wool shirt and jeans. "I've got an extra slicker too. I'm here for another week. Just drop them off when you're passing by again." The warden's name was Sven, a Swede with a gentle, lilting accent. The small cabin was immaculate. A bed, neatly made, ran along one side. Shelves stacked with provisions and books lined the walls. We sat at a little table, drinking coffee and talking of horses, the lousy weather, a difficult bear that Sven had had to shoot, and the increasing numbers of hikers making their way into the park. Warmth crept through me. I wasn't used to drinking a lot of coffee, but with the friendly talk, and at Sven's insistence, I had cup after cup of the hot fluid, heavily laced with Eagle Brand Sweetened Condensed milk. By the time I realized I had better get going, I was enjoying a euphoric sense of well-being.

In dry clothes and Sven's slicker, I went with him to the corral. It was still sleeting, but that didn't bother me now. I felt like I would never be cold again. Sven politely assumed the role of offsider as we repacked Goldie. He made admiring noises at my carefully executed diamond hitch. "Sure wish I could get my assistants to do one of them properly," he said. My heart swelled. Clearly Sven was a man of exceptional judgment.

On the trail again, snug, warm, and dry, the world seemed a different, better place. Even the horses stepped up smartly as I performed my rendition of Dylan's *The Times They Are A' Changin'* at the top of my lungs. A coyote puppy gazed down sweetly at me from the crest of a small ridge. Dark was falling now, and the rain began to ease off. Our corral was a little outside Tent City, and I happily drove the last horse in, unpacked and hobbled them. This was my favourite moment. Freed at last of their loads and let out to graze in the valley for the night, they would shake themselves and often have a roll. "A horse is worth a hundred dollars every time it rolls," Tom had told me. I wasn't exactly sure what this meant, but I knew he was right.

It was night by the time I made my way into the huge cook tent. This was staffed by charming, elderly women, also on contract to the Alpine Club. At least half a dozen wood-fired cookstoves made up a centre aisle. On either side, pack boxes served as seats in front of long plywood tables on which massive platters of roast beef, hash-brown potatoes, and hot corn bread were laid out. As the only packer there that night, I enjoyed a privileged position, and second helpings were piled onto my plate. Outside, around an enormous bonfire, alpiners awaited their promised magic show. In the light of the fire, warm and well-fed, I performed eagerly. Everything went right. Coins appeared and disappeared; ropes were cut and restored. I stole watches easily and got a tumultuous response as my dupes stared down at their empty wrists when I asked for the time. The final applause after the Chinese Linking Rings was genuine and generous. When it was over, I felt like the most popular and important person in the camp.

In high spirits, I said my goodnights and set out to find the lean-to. Away from the fire, I noticed for the first time how black the night had become. The stars and moon had disappeared, and only the faintest outlines of the mountains gave any hint of the right direction. Blindly, I groped my way through the brush. A creek ran past the front of the lean-to. Not wanting to tumble into it, I carefully felt my way forward with every step. At last, I felt the side of the bank with an outstretched foot. Stepping back a few paces, I poised myself for a run and a jump. I had no intention of getting wet again. Taking a leap far larger than was necessary, I soared through the night air—and smacked into something large, warm, and firm. There was a terrified whinny as I felt the whoosh of hooves graze my ribs. In terror, I recoiled straight into the icy stream and lay there, water damming up against my body and flowing over me, trying to recover my wits. I had jumped straight into Ollie's big black rump.

CHAPTER 4:
OTTAWA (1963-1964)

It had been an endless summer. I was ready to be a cowboy for the rest of my life and returned reluctantly to Grade 12 in suburban Ottawa. My language was now peppered with expletives and double negatives. I remember trying to relate one of Tom's stories in an English assignment, attempting to capture his version of the exploits of a certain Luke Green. This man, a prominent Jasper character, had been blessed with seven daughters. Under pressure from his girls, he bought a dog, "the stupidest damn dog you ever did see." Well, one day while Luke was in the supermarket, the dog, frantic from waiting outside for Luke's return, took advantage of a slow-moving, incoming customer to dart through the open door. First, he headed straight for the display of Tide that had just been fixed up into a kind of pyramid. Cocking his leg, he peed all over one side of it. Then, for good measure, he barrel-assed to the other side and peed there too. Soon Luke, the manager, and a dozen random shoppers were after him. The dog thought it was a game. Running at

Magic Travels

full speed under the counters of goods, he knocked down a full stand of fruit. The shop was in chaos. Every time there was a lull in the chase, he cocked his leg on whatever was handy. Finally, he was cornered and pitched out. Luke collapsed in exhaustion on the cashier's counter. "You know," he said, "you know... I've raised seven girls... seven! Never had no trouble... 'til I got that damn dog." Perhaps you had to hear the story from Tom himself. It didn't pass muster with the English teacher, nor did I pass the assignment.

But before leaving Luke altogether, one last anecdote comes to mind. Not surprisingly, Luke was an active member of the Parent Teacher Association. One year, he was even elected chairman. As I heard it, he did fine, apart from getting a few of his mords wixed up. I was present when Tom and Fay returned from one of the meetings. Tom was in fits of laughter. The issue of obtaining a substitute teacher while Miss Pratt was away on sick leave had come up. "Well, I don't see no problem," announced Luke. "Why can't Mrs. Baxter prostitute for a while in Miss Pratt's place?"

Now, there was a full school year to endure before I could return to Jasper. It wasn't that I was a bad student, but I was no genius either. I dreamed of cutting and running, but my father made it abundantly clear that continuing on to university was not a choice; it was a given. I yearned for travel, excitement, adventure. By comparison with the cowboy life, school seemed unutterably dull.

But there were amusements.

From the days of summer camp, I had always had a love for sailing. On Dow's Lake, only five minutes from the classroom, I had a little Flying Junior dinghy (named Pippin after one of the hobbits in Lord of the Rings). In class, whenever the wind was blowing, an impressive row of tall elm trees rustled and swayed against the side of the school. The sight and sound of those trees drew me insistently away from the stuffy classroom, and as soon as the bell rang, I was out the door and down to the boat. In the wonderfully shabby little boathouse, there was a world map on the wall with a printed caption below: PLACES YOU CAN GO FROM DOW'S LAKE. I took that phrase to heart when I realized with a thrill that it was true. Why, you could sail across to the far side of the lake, turn left, and head down the Rideau Canal. From there, it was a flight of eight locks through to the Ottawa River, turn right to Montreal and left again at the Saint Lawrence. After that, the Atlantic Ocean and the whole world would be yours. Or, of course, you could go the other way. Cross the lake and turn right, stay on the canal to Kingston, then left into the Saint Lawrence, and again the world was yours. The idea set my blood stirring.

The farthest I ever got from Dow's Lake in Pippin was during an attempt to impress Rachel. Having admired her in the hallways of Glebe between classes, I had aspirations that she would see the light and be my new girlfriend. When I summoned my courage to ask her out for a sail, she seemed thrilled by the idea. We set out on a windy Saturday afternoon. Crossing the lake, we did indeed turn left and headed down the canal towards the parliament buildings. The wind was behind us, and we went rapidly under the Bronson and Bank Street bridges. A couple of miles further on, we found ourselves blocked. The mast was too high to get under Pretoria Bridge. As we hauled in the sails to bring the boat round, I became aware that our tailwind was much stronger than I'd realized. At that moment, a sudden squall hit and over we went. That was not an uncommon experience but being with Rachel was. Righting the Flying Junior and getting underway again was something I was well acquainted with, but this time not only did I have a prospective girlfriend with me, we were already rather too close to Pretoria Bridge for comfort. To put it plainly, we were in that position sailors fear—namely, in trouble off a lee shore. Not only that, but the wind was not content with just a knock-down. The newly exposed hull enabled the wind to roll the boat completely over. We were turned turtle in other words.

The centreboard now stuck straight up into the air, and just to make things more difficult, the mast became stuck in the canal's muddy bottom. That was good as far as anchoring the boat was concerned. At least, it couldn't be blown against the bridge. It wasn't so good in that—well, what should we do now?

I had the feeling that Rachel wasn't too impressed as I instructed her to hang onto the upside-down rudder. She was a strong swimmer, however, and wasn't panicking. We were also, as I realize now in writing this, replicating the spectacle that had occurred on Pyramid Lake Road, at least when it came to halting traffic. Cars on Queen Elizabeth Drive, Pretoria Bridge, and Colonel Bye Drive were slowing down, some even stopping. An audience began to form at the rails on both sides of the narrow canal. There was some shouting. Did we need help? Should the police be called? Embarrassed, I shouted back that everything was okay. Knowing, with the mainsail still up, that the water resistance against it would make righting the boat impossible, I told Rachel what I planned to do, then dived under and came up into the air trapped beneath the hull. Finding the cleat in the darkness, I undid the halyard and pulled the upside-down mainsail upwards off the mast. I was aware that the small amount of air I was breathing was quickly being depleted, and it was a relief to dive back under the gunwale to the surface.

We were in luck. The wind had died down somewhat. I told Rachel to put all her weight on the transom while I did my best to jerk the bow upwards. With relief, I felt the mast free itself, and in short order, we were able to clamber up on one side, grab the top of the centreboard, and use our combined weight as leverage to rotate the hull back to an upright position. After that, it was easy to scramble aboard since the boat was wallowing so low in the water. I soon had the main up again, and we started moving forward, as close-hauled as we could go, to get away from the bridge. As we gained speed, the water began to drain out through the transom scuppers. We actually got a round of applause from the shoreline as we finally got away, but our troubles were not over.

We faced a strong headwind, and the canal could get as narrow as twenty-five metres, so we were forced to return in a continuous series of short tacks. It was a long and arduous journey back to Dow's Lake. Each tack took only moments before we had to come about again, and each coming-about had to be done exceptionally quickly to avoid losing precious upwind distance in the interval between carrying out the tack and getting the boat up to speed again. It fell to Rachel to handle the jib sheets. With each tack, the leeward sheet had to be released at exactly the moment the wind caught the other side of the sail. Amid much wild flapping, Rachel had to pull in the other sheet and, at the same time, shift her weight to the windward side of the boat. It seemed to take hundreds of tacks to make it back to Dow's Lake. I sensed that this wasn't the idyllic Saturday afternoon sail she had been expecting.

But I was still taken aback when, in a brusque exchange the following day, Rachel waved her blistered hands at me and blurted out, without preamble, "Patrick, I am so sorry, but my parents have told me not to see you again. We are Jewish, you see, and I'm only allowed to have Jewish boyfriends." I consoled myself by believing her and imagined that she was heartbroken to have to tell me this. It was perfectly understandable that her God would find me unacceptable. I had already, as a child, had a near brush with the Presbyterian God and later with the Anglican. It would be implausible now to claim that, surprise, surprise, I actually have Jewish ancestors and was seriously thinking that the Jewish God was probably the best and that I'd already been thinking of converting. Sadly, I accepted my fate. The future with Rachel that I had been fantasizing about was not to be.

∞

Magic Travels

Growing up in Ottawa in the 1950s, I had plenty of exposure to God. For some reason, at the age of six, I was shunted off to Sunday school at St. Giles Presbyterian Church at the corner of Bank St. and First Avenue. Why Presbyterian? I have no idea. My mother, Phyllis, had been brought up in the Church of England and its unobtrusive style. If she was a believer, her faith was low-key and inconspicuous to say the least. She certainly didn't go to church. Her life with Digby, my father, would have made continued belief in a deity unlikely. He had started out as a paleontologist, studying the evolution of animals recorded in fossils preserved in sedimentary rocks. It requires the most convoluted mental gymnastics to accept the reality of evolution through eons of geological time and still believe in a creator, particularly one who did all the work in six days.

My father may not have been a dyed-in-the-wool atheist, but he was certainly agnostic. Coming home one Sunday morning, I related the teachings I had just received from the Reverend John Logan-Vencta who I later learned had a reputation of some renown. Rev. Logan-Vencta had that morning told us that when little birds land on electrical wires they are not electrocuted despite the large voltages of current flowing through them because God, in His infinite mercy, protects his precious creatures, especially little birds. We had then sung a hymn expressing our thanks for His goodness. My father, smiling, explained that electricity always looks for a way to get to the ground via the path of least resistance. Since the little birds are not touching the ground or anything in contact with the ground, the electricity stays in the power line. Even you and I, he told me, could hang from a power line with no ill effect but would die if we also touched the ground. I asked him if he did not believe in God. "I can find no evidence for the existence of a god" was his reply.

I did not ask why, in that case, I was being sent to listen to the Rev. Logan-Vencta's fairy tales. The reasons I was sent to Sunday school remain one of those mysteries of childhood. Perhaps it was simply to help me make friends, or perhaps to learn something about religion, so I could decide for myself. It might simply have been a means of having their Sunday mornings free for themselves. Whatever the reasons, I was never encouraged either to believe or disbelieve, but I did have a lot of science instilled into me from a young age, which no doubt simplified my path away from religion.

After learning the truth about birds on high tension wires, I began finding St. Giles hard to take seriously. My friend David Sinclair persuaded me to come over to St. Matthew's, the Anglican Church just across the street. David was a choirboy, and I liked the idea of being one too. It wasn't quite so easily done. I remember standing before a committee of elderly women, feeling intimidated. They towered over me and seemed uncertain about whether I should be allowed into St. Matthew's. As the interrogation proceeded, they kept glancing doubtfully at each other. I finally gleaned the problem. They knew I had defected from their rival across the street and wondered where my true loyalties lay. Eventually, Reverend Osborne had to be called for. He turned out to be a lovely, unsuspicious, welcoming man, and I took to him immediately. Of course, I could join the church, he said. He would set up my audition for the choir. The elderly women looked nonplussed but evidently knew better than to object.

I don't regret my time at St. Matthew's, but when I picture myself in a white ruff collar, red cassock, and white surplice, my days as a choirboy can seem bizarre. Choir practice was twice a week, with performances at morning and evening services on Sunday. When we were lucky, we sang at weddings for a silver dollar each, quickly spent at the Mirror Grill, a greasy spoon almost next door on Bank Street. Hamburgers were twenty-five cents. The choirmaster, Gerald Wheeler, was an exceptional catch for the church, having been assistant organist at Saint Paul's Cathedral in London. With his help, I got to know and even love a considerable amount of classic liturgical music. I also got to know the Anglican service in detail, especially when to get up or down from a kneeling position. I listened to the prayers and sermons, of course, but they were like water off a duck's back. I never actively decided to disbelieve. From the start, it just seemed absurd. Occasionally, while on my knees with my head buried piously in the folds of my surplice, I would try to pray but quickly felt uncomfortable, hypocritical, or just plain ridiculous. I certainly didn't fear the wrath of God. It was as if the whole concept of belief was foreign to me. I didn't begrudge others their faith. It just wasn't for me.

∞

Amid choir duties, piano lessons, and the school ski team (downhill, slalom, jumping, and cross-country), winter term at Glebe still allowed time for a mandatory trip to Mount Marcy. Located in the High Peaks Region of the Adirondacks near Lake Placid, Mount Marcy, at 5,300 feet, is the highest mountain in New York State. During Christmas holidays, Guy Sprung and I, sometimes with several others from our circle of friends, packed up rucksacks, cross-country skis, and snowshoes and set out to hitchhike to Lake Placid, where trails that led to Mount Marcy were close by. The route was simple enough, to begin with anyway: Bank Street south to Morrisburg on the Saint Lawrence, then the famous Highway "Four-Oh-One" east to Cornwall. There, the border had to be crossed—always a little dicey when you're a hitchhiker—followed by a variety of smaller roads southeast to Lake Placid. Total distance: 146 miles. Because it was difficult to get a ride for more than one person at a time, not to mention the amount of gear required for skiing and camping in winter, we always hitchhiked separately. Now that we rely on smartphones and GPS to find our way anywhere, it seems extraordinary that we managed to meet up again, but I don't remember anyone ever getting lost.

The availability of lifts, however, was always a problem. The day I remember in 1963 was no exception. December on an empty, windswept highway on the outskirts of Ottawa is cold. With a pile of gear at the side of the road, the wind howling, and sheets of snow blowing into dunes across the asphalt, there was no sign of traffic whatsoever. I had been standing for what felt like hours and was desperately cold when a pair of headlights finally appeared from around the bend. I hurriedly stuck out a mittened thumb and added little imperative waves with my arm. Staring hard through the blowing snow at the shape in the driver's seat, I composed an expression of pitiful entreaty. Seconds later, I dropped my arm and turned away. What was the point? As it drew closer, the vehicle revealed itself as a large, rather fancy, black hearse. As I expected, it passed without slowing. I turned my attention back to the highway for a sign of any further traffic. Unexpectedly, the sound of a horn competed with the wind. I turned and looked disbelievingly at the hearse, now stopped about fifty metres further down at the side of the road. Its glowing taillights could not have looked more inviting.

I gathered my gear and hurried toward it. The passenger door was already ajar, and before I could even start my thank yous, a voice instructed me to chuck my things in the back. Opening the large back door, I was flabbergasted to find the space almost completely taken up with a coffin. "Just shove your stuff on top," the voice continued. "She won't mind." Gingerly, I placed skis, poles, and snowshoes on the polished lid of the casket and then my heavy, bulky rucksack. In the front seat, I found a small tubby man with a grey, heavily wrinkled face. He was dressed in pinstripe trousers that were somewhat too tight, shiny black pull-on shoes, and a morning coat. A black top hat rested in the space between the seats. Incongruously, a roll-your-own cigarette smouldered between his lips. I thought of illustrations of Mr. Micawber in David Copperfield, the novel we were taking in English that year.

Magic Travels

Trying desperately to recover from the shock of finding myself inside a hearse with a corpse in the back, I put on my calmest, coolest demeanour. It was a normal, everyday experience, I told myself, happens all the time. Besides, it was blissfully warm inside. I thanked Mr. Micawber profusely, but he paid no attention. He was a non-stop talker and seemed to care little about what was said in return. His language was riddled with expletives. "There was a fuck-up at the Home, and I had to leave Cornwall early this morning to get the lady from Ottawa. One fucking screw-up, let me tell you. Her service was changed back to Cornwall, so I'm in a fucking hurry." The last detail was evidently true. Despite the snowstorm and the perilous road conditions, his speedometer hovered steadily between eighty and ninety (mph in those days). "Cornwall?" I said, not believing my luck. "That's exactly where I want to catch the bridge over to the States."

"Sure, sure, no fucking problem. I go right past the goddamn intersection."

We soon turned onto the 401. It was just as bleak if not worse. The snow was blowing hard, but there was virtually no traffic. In a flurry of cuss words, my benefactor floored it, and we shot off into the blizzard. Shortly after, we slowed. Mr. Micawber was frowning at the dials. The hearse slowed to a crawl, and we pulled onto the verge and stopped. "Jesus fucking Christ! The boys at the Home told me they had filled her up. Son of a bitch!" Miraculously, he produced from nowhere a five-gallon jerry can, put on his top hat, and told me to wait. "I won't be fucking long if I can help it." Out he climbed and disappeared into the snow. So, there I was, sitting alone, marooned on the 401 in a snowstorm, in a hearse with a dead lady in the back, and all my skiing and camping gear on top of her. This, I thought, cannot be normal.

Where exactly he went and how he got five gallons of gas into the jerry can will remain a mystery forever. Despite his love of talking, he never told me. I suppose he must have found a nearby farmhouse since he was back within half an hour. He was so cold and screaming mad when he returned that I didn't dare ask. We soon found a gas station and filled up. And he did, as promised, drop me off exactly where I needed to be to continue my journey. I could hardly complain.

Incredibly, only a month or two later, I found myself once again in a speeding hearse. The Glebe ski team had come down from Ottawa to race at Mount Washington, a resort near Lake Placid. In a practice run on the downhill course, I caught an edge and fell hard on icy snow. In severe pain, I was strapped into a rescue toboggan and carted off by the ski patrol flat on my back, unable to see what was happening, and loaded headfirst into an ambulance. So, I was surprised to look up and see a large velveteen cross embroidered into the upholstery of the ceiling. I felt the vehicle take off at speed, siren blaring, and within minutes we arrived at a hospital, and I was able to confirm that the small town of Lake Placid used a hearse to double as an ambulance. Of course, the obvious joke crossed my mind: I had been lucky the attendants hadn't got their destinations mixed up. I was alive, but with a couple of broken ribs that finished my skiing for that winter. And I'm happy to report that to date (2021) I have never been near a hearse again.

∞

The trip to Mount Marcy was particularly miserable that year. With the extreme cold that winter, the place was deserted. We plowed our way through unbroken trails closer to the foot of the mountain where there was a frozen lake and a number of framed lean-tos set up for campers. We soon got a large bonfire burning merrily, providing some warmth while we set up our camping gear inside one of the lean-tos. During the night, it was so cold that I awoke to find John shivering beside me in the same sleeping bag. Under the circumstances, I could hardly object. We put his sleeping bag over both of us, which helped some more. The next day, we did the climb to the top of Mount Marcy. The exertion produced some warmth, but at the top, it was too cold and windy to stay and admire the view. We knew the trip back down on our slender cross-country skis in unbroken soft snow would be tough. And it was. It took many long traverses across the open mountainside above the treeline, and falls were frequent. It was even more difficult in the trees, where steering was more luck than skill. The trick was to go fast enough to surf on top of the snow and rely on jump turns to steer. But a fall at each turn was more common than not and getting back up again was a struggle in the deep snow. We were utterly exhausted when we arrived back at camp. That night we heated small boulders in the bonfire and took them into our sleeping bags. Their warmth lasted long enough to get to sleep.

The next morning I went for a walk alone. On the way back, I decided to take a shortcut across the frozen lake. As I approached the shoreline, I must have walked over an upwelling spring. Without warning, the ice collapsed beneath my feet, and I plummeted through the icy water, dragged downwards by my heavy winter clothing. It is impossible to describe the incredible wave of relief when my feet hit bottom just as my chin came in contact with the water. How I got out I'll never know. I do know that the body can draw on unrealized potential at moments like that. I only remember the thin ice repeatedly breaking as I tried to pull myself up, but I couldn't have been submerged for long. The next thing I recall is running toward camp and shouting to get the fire going. I peeled off all my clothes and was able to scrounge enough garments from the three of us to keep the worst of the cold at bay. The clothes I'd been wearing froze solid immediately, but with careful tending around the bonfire, they were dry again by the end of the day.

We skied and hitchhiked back to Lake Placid the next day. The temperature kept dropping, and as night fell, we realized it would soon be too cold to travel further. We had heard that sometimes the police would let needy people sleep in their jail. Rather desperate by this point, we made our way to the station off the main street. A lone constable was lounging in his chair behind the front desk. He scarcely looked up from his magazine when we asked if we could stay overnight in the cells. Gesturing toward a heavy door off one side of the office, he said, "Through there, but you'll have to leave if we start to fill up." Gratefully, we filed into an unlocked, barred cubicle holding two iron bunk beds. Thin grubby mattresses and a naked toilet completed the furnishings. There was a smell of disinfectant overlying odours that were better ignored. But it was warm, and we all slept well until, "All up! You've got to leave now!" The voice was strong enough to rattle the bars. Hurriedly, we packed up, thanking the officer in charge as we stepped into the cold dark of early morning. It was six o'clock. Time to separate and begin the long hitchhike back to Ottawa. And that was the one and only time in my life I have experienced the inside of a jail cell.

CHAPTER 5:
JASPER (1964)

Guy Sprung had written to me the previous summer in Jasper to say that he would like to hitchhike out to visit me. Remembering the length of the train journey (three days and two nights), I advised against it. "Don't do it. You have no idea of the distance." Going by train had overwhelmed me with a sense of how big Canada actually is. It had seemed to take forever just to get out of Ontario. But Guy ignored my advice and hitchhiked out, even beating the train by a day, and I spent a pleasant few days introducing him to cowboying.

Simple pride now meant that I couldn't possibly take the train for my next summer with Tom McCready. The last days of school ticked slowly by while I hopped about in excitement and impatience. Immediately after writing my last Grade 12 exam, my mother drove me out to a carefully selected spot on Carling Avenue. I wanted a place where cars would see me clearly and be able to pull over easily and safely. From there, the route west was long but uncomplicated: five miles to the Trans-Canada Highway, then 2,160 miles straight west to Lake Louise, Alberta, where I would connect with Highway 93, the famous Banff-Jasper Highway and travel north another 385 miles.

By nightfall, I had made only the first 180 miles. I was standing on a deserted road on the far side of Mattawa, thinking the day was over and looking around for a place to bed down. I spied some woods ahead and walked along the verge toward them with my back to oncoming traffic. Luckily, I kept my thumb stuck out into the road, and my optimism paid off. A half-ton truck passed me, swerved onto the verge, and stopped. "Jump in the back," the driver shouted. He appeared to be in a rush, and there was a passenger already sitting in the only available seat beside him. I chucked my gear into the empty cargo space behind the cab and hopped over the tailgate to join it.

For a while, it was pleasant in the fresh air and knowing the miles were passing again. We had entered the vastness of northern Ontario, and I began to get worried that I didn't actually know the driver's destination. As long as we stayed on the Trans-Canada, all was well, but I would have to bang on the roof if I became aware of a change in course. As night fell, it grew colder in my open cargo bed. I began to feel curiously isolated, alone in the empty back, as we hurtled through the darkness. I could see the two men through the back window in the cab, their cigarettes glowing in the dark. They didn't seem to be talking much. It grew uncomfortable on the hard floor. Finally, stiff with cold, I unpacked my sleeping bag and crawled in, anxiously wondering where our destination might be, and whether I should let myself fall asleep.

"You wanna get in the front?" The voice awoke me from the depths. Neon lights pierced my foggy eyes. We had stopped in front of an all-night café somewhere west of Sudbury. It was two o'clock in the morning. This was the end of the line for the passenger, who had apparently been hitching as well. I climbed out of my sleeping bag, feeling somewhat embarrassed. "Where are you headed?" I asked. "Revelstoke," came the reply. Revelstoke! Had I heard right? That was 260 miles west of Lake Louise. If he dropped me off at Lake Louise, I would have made nearly the whole distance from Mattawa in one lift. One thousand, nine hundred and seventy-five miles! Would that be a hitchhiking record? "I'm trying to get to Jasper, so that would be fantastic!" I said, hoping that the prospect of such a long distance in my company wouldn't deter him. "No problem. Do you drive? I want to go straight through. I could let you off at the Banff-Jasper Highway." I had taken my driver's licence only the week before in preparation for my second summer with Tom. Driving was still a thrilling novelty, but recklessly I suppose, I saw no reason to tell my new companion. "Sure," I said.

"Well, can you take over now? I'm about all in." He handed me the keys, climbed into the passenger's side, curled himself up, and closed his eyes.

Excited, I started the pickup, turned on the lights, pulled out into the empty road, and accelerated to sixty. The first approaching car violently blinked its lights at me. I realized that mine were on high beam, but I couldn't find the floor switch right away in a strange vehicle in the dark. Seconds later, a siren sounded behind me. A glance in the mirror showed a myriad of bright flashing lights. A squad car pulled alongside, the policeman making gestures that left no room for misinterpretation. I pulled over onto the shoulder and anxiously climbed out. A Mountie was already getting out of his car.

"Licence," he said.

"Yes... yes, sir," I stammered, fumbling in my wallet in the glare of his flashlight. "What's the matter? Is anything wrong?"

"Name." I told him. He carefully confirmed it from my licence. Apparently not satisfied, he continued. "Address," he intoned. I told him and seemed to get full marks on that too. "Okay," he said. "Where are you going?"

"Going west, officer ... sir. Heading to Sault Ste. Marie and all the way to... Alberta... sir."

I sensed that he didn't like that answer.

"Let me smell your breath." I blew a lungful at him. "Do you realize that you didn't dim your lights as you drove past me just now?"

"Yes sir, officer, sir... I'm very sorry, it won't happen again... sir."

"Let me get this straight. Where did you say you were going?"

"Like I told you," I bleated, "all the way west... sir." He was looking grimly at me now.

"Do you mind telling me then how come you're heading east toward Sudbury?"

"Oh my god... really?" In a flash, I realized that, having awoken in the dark, I hadn't even seen which side of the highway the café was on. I had simply assumed the truck was pointing in the right direction. A right fool I was looking now. Obsequiously, I tried to explain my mistake, but I suppose he decided there must be bigger fish than me out on the tarmac ocean. With a bored look, he let me go with a verbal warning. I climbed back into the cab and before turning the vehicle around made sure I knew where the high/low beam switch was. My friend was still asleep. I couldn't help but feel grateful for the cop who saved me from driving an unknown number of miles the wrong way.

I drove through the night. At daybreak, my partner awoke. He was a middle-aged man with sandy hair and a big belly. On the backs of his right hand fingers the words, I LOVE YOU MEG, were tattooed. The fingers of the other hand were inscribed, I LOVE YOU AM.

"There wasn't room for Anne-Marie," he explained.

"Girlfriends?"

"Wives," he replied sadly.

Despite his tattoos, Ned, for that was his name, insisted that neither wife knew of the other. He was immensely proud of this. But he was feeling some financial strain on his elevator mechanic's salary. This was the reason for getting as quickly as possible to Revelstoke. He had, he excitedly told me, found *true love*.

"And she's rich!"

⁂

Two and half days later, my last lift dropped me in the middle of Jasper. I walked the few blocks to the barn. "Jeeeeeesus, you sure must be lucky at hitchhiking," Bill said as he pumped my arm. He was right. Like Guy the summer before, I had beaten the train by a day, and it hadn't cost me anything except sleep. I was soon established in the loft with its familiar smells of leather and canvas. Tom had a more regular season in store this summer. Instead of dealing with the Alpine Club, we were going to be busy with his summer fishing camp at the north end of the Amethyst Lakes in Tonquin Valley.

I had arrived just in time for the spring roundup. This meant that all Tom's horses, having spent the winter fending for themselves in various range areas, had to be found and brought into the corrals around the barn. My tracking and riding skills were going to be seriously tested. I had to find the reluctant herds, then drive them through the bush back to Jasper, often at a wild gallop. Initially, I rode with Tom, who took pains to explain how to read horse tracks to tell how old they were and whether they were worth following. When a clean, fresh track was found, Tom could often tell from its pattern which horse it belonged to and which one needed a new shoe, even on which leg. Because the herds tended to band together in consistent groups, identifying only one horse could lead to the correct identification of the herd we were looking for. Often, a bell had been put around the neck of the dominant horse in each herd, and tracking was sometimes a matter of just stopping and listening.

As I became more experienced, Tom began to let me ride Abby, the best horse for a successful search. Alone with Abby, I had the enormous pleasure of no longer being a dude or so I convinced myself. Together, he and I searched hundreds of square miles for tracks, all the while listening for the sound of a bell. Abby knew what we were doing and took an active role. He would hold himself still, ears twitching, nostrils quivering, concentrating all his senses on finding a herd. When he did, it was foolish not to let him have his head even if you thought you knew better. He would set off confidently, nose to the ground like a blood hound, never wavering, and even if it took several hours of riding, would take you to them.

That's when the fun really began. Tom's horses were definitely not the pony barn kind. After spending an entire winter roaming freely, they were essentially wild again by the time each spring came round. Such horses were considered vastly superior to their pony barn counterparts who were pampered in stables all year-round. They lived as much as ten years longer and were known to be far more sure-footed on steep trails and especially crossing rivers, where hooves struggled to maintain balance on unseen and unstable boulder bottoms. But they were also… well, wild. They did not take kindly to resuming the life of bondage that they seemed to know awaited them if caught.

The trick was not to startle the horses. It took little to set them off at high speed in the wrong direction. The best strategy was to circle quietly around them, giving them time to get used to the arrival of horse and rider, and at the same time, make a count to ensure that none got left behind in the ensuing roundup. When properly positioned, they could then be gently nudged in the right direction. With encouraging shouts and a few flicks of your reins on a flank or two, it was sometimes possible to induce an orderly progress from the bush to the nearest trail back to Jasper. More often than not, the process was far from orderly. All it took was for one of them to be spooked and break into a gallop. The rest would instantly follow suit. Then the mad, exhilarating chase began.

If they changed direction, I had to race past them, reach the leaders, and steer my mount across their path to turn the herd onto the right course. When I was lucky enough to be riding Abby, this was a positive pleasure. He knew precisely what was expected of him, and his speed was second to none. After the initial hysteria subsided, it became clear that the herd remembered the route well. Once resigned to another summer in captivity, they usually maintained a steady speed back to the corrals. Perhaps a memory of free food, especially their beloved oats, was enough to persuade them that being put back to work was not so bad.

Next came the shoeing. As I was soon to learn, wearing a pair of thick chaps was essential. Tom had an assortment of worn and ominously scarred pairs, one of which he tossed over to me. He selected Sally, a large good-natured mare, for my first lesson. Leading her into the barn, Tom showed me how to relax the animal by talking gently to her and stroking her neck before running my hand down her front leg to her ankle or, more correctly, her fetlock. With a little encouragement, a good horse might allow you to lift its foot by the fetlock to get the hoof off the ground. A not-so-good horse could usually be persuaded with a few taps on the fetlock with the shoeing hammer. Bent over with your back to the horse, you could now pull the hoof up and hold it between your legs. Firmly. It doesn't take much imagination to understand the consequence for your most vulnerable parts if a horse decides it's had enough of standing on three legs and kicks out in anger.

I had no sooner got Sally's leg between my own when I felt her hot breath and her nose pressing into my neck. In a panic, I released the leg and sprang away. "That horse tried to bite me!" I shouted. Tom, and Bill who had come in while I was bent over Sally's hoof, nearly fell over laughing.

"Sally's only giving you a kiss," explained Tom.

Mortified, I realized at once that he was right. There had been no aggression in Sally's nuzzle. Absent the surprise, it had felt rather pleasant in fact. Too chagrined to say anything, I resumed my task, and the nuzzling soon turned into a bonding experience. This was just one, but certainly the most agreeable, of many behaviours that horses can inflict on the person attempting to nail metal shoes onto their hooves. One of the worst was *the lean*. This happened when a horse decided that, since you were responsible for lifting one of its legs off the ground, it might as well use you as the support it was now missing. The result was that you became more and more aware of a great weight pressing on your body as you struggled to keep the hoof in place between your legs. Eventually, the weight became too much to bear, and you would feel your back giving way. Now, the problem was to get out from the position you were in. The only way was to simultaneously drop the leg and leap clear as fast as possible while the horse desperately tried to recover its balance. I never saw a horse fall right over, but it was sometimes close.

It was important that all the necessary tools were within arm's reach on the floor before placing the front hoof between your legs. These included a hoof pick to clean out the foot before getting to work; a shoe puller and nail puller if there was an old shoe needing to be removed; nippers, basically a giant pair of nail clippers, which were needed to trim around the hoof wall until the foot was the right length; and various knives to pare away excess sole and remove loose, dead frog (a triangular-shaped spongy organ containing nerves to sense the ground) so that healthy tissue could breathe. Of great importance is the rasp, a huge nail file to even out the trim and level the hoof. The rasp also gets used at the end of a shoe job to smooth out nails and make sure the edges of the hoof exactly meet the edges of the horseshoe. The process is much the same for the back hooves except that rather than bringing the hoof up between your legs you rest it on top of your thigh.

It is the rasp that makes me reflect on the work of shoeing with a certain amount of discomfort. While I never doubted that Tom loved his horses, they were working animals and at least half wild. And there were a lot of them. A difficult horse could not be tolerated, and training could not afford to be gentle and slow. If twenty horses needed shoeing that day, the one that decided to kick, bite, or otherwise make things difficult received an emphatic beating. The rasp, long and heavy, with a savagely roughened surface, and always near to hand, was ideal for the job. One blow, let alone several, must have been seriously painful, but both Tom and Bill never hesitated to let fly whenever they considered it necessary. I naturally followed suit. The accompanying language was loud, colourful, and dramatically creative, and I picked that up too.

When Tom won a contract to shoe all the pony barn horses at Jasper Park Lodge, I understood why he made us carry out the job in a remote clearing in the bush far removed from the eyes (and ears) of the management and guests. I reconciled myself to the cruelty of our method with the thought that it got the job done as quickly as possible. I don't remember ever seeing blood, and the horses never seemed obviously distressed once the ordeal of having eight nails hammered into each of their four hooves was over.

This brings me to the most difficult part of the work: the actual nailing. Once the hoof was properly cleaned, trimmed, and levelled, a shoe had to be found from the huge supply that was lined up on racks in a nearby shed. The variety of shapes and ages was enormous, and it wasn't easy to find a shoe that would fit the hoof well. We did only cold shoeing. There was no forge to heat and hammer a shoe to the right shape. If a previously used shoe was chosen, it usually had bends in it. I had to hammer these out and make it perfectly level again. It was also important to get the weights similar for each pair of hooves.

Horseshoe nails are unique. They have a bevelled point that enables the nail to bend outwards and emerge out the side as it is driven into the hoof. Before beginning, all eight nails need to be loaded between the lips ready to use. The nail should exit about a thumb's width above the hoof bottom. If you tap too lightly, the nail will come out low. If you use harder blows, the nail will come out too high. Nails at the front of the hoof will come out lower than nails along the sides. Getting the nail pitch and hammer blows just right takes practice. For the horse, the process is not painful as long as the nail emerges through the side of the hoof. If not, the nail will penetrate the nerves of the foot (like a dentist's drill hitting an unfrozen nerve) and can do considerable damage. And the horse will let you know it, possibly resulting in considerable damage to you.

Even when a nail is hammered in correctly, no horse is happy about it. The jarring caused by each blow of the hammer is unsettling, and when the nail emerges from the side of the hoof, the naked point can be dangerous. If the horse chooses that moment to pull its leg back, the nail can be dragged into your inner thigh (or worse). Gashes made by nail points were clearly visible on the scarred chaps Tom had given me to wear. I had to act quickly, using the hammer to knock the sharp end of the nail away from the slope of the hoof and then, with a twist of the wrist, the hammer's specially designed claw breaks the sharp end off, leaving only a tiny protruding stub of metal.

When all eight nails are quickly hit home with few misses of the hammer on the nail head, the horse seems to know that the ordeal will soon be over and realizes that cooperation is helpful. It is part of that unspoken communication that can take place between horse and human, a brilliant experience when it happens. All that remains are some finishing touches. The hoof is brought forward and supported on the knee. The rasp is used to file a small notch on the underside of each snubbed-off nail point, and then, using specially designed alligator tongs, each nail stub is clenched downwards into the notch, leaving no sharp protrusions from the hoof. The rasp can then be used to tidy up the nail end and hoof.

Unquestionably, shoeing horses is one of the hardest jobs I have done in my life. It is backbreaking, occasionally painful, and hugely filthy. Most horses, being nervous, release their urine and droppings all over the tools you are working with. A heavy grime accumulates over your skin and clothes, and by the end of a day, you reek of horse. I wouldn't have missed the experience for the world. Later, working on my master's degree at Calgary, I made extra money shoeing horses. I charged a dollar a foot and took an hour for four of them.

∞

The work this summer was different. There was plenty of riding and packing to do, but without the demanding schedules of the Alpine Club, the whole tenor of Tom's operation was gentler, calmer, and more varied. The trail to the fishing camp took a different route into the Tonquin, following Portal Creek Valley, crossing over Maccarib Pass, and back down to the Amethyst Lakes. The views were just as spectacular, and the river crossings just as frequent and tricky. Maccarib Pass, at about 7,200 feet, was well above the treeline where the wide open spaces gave rise to meadows. Red and yellow paintbrushes, aster, goldenrod, and innumerable other wildflowers provided stunning displays. Marmots, with their distinctive whistles, bears, caribou, and porcupine were common. Tom's clients, often family groups of three or four, were from all over the country, and he often made me responsible for all the packing and guiding for their holiday. Amethyst Lake was plentifully stocked with rainbow trout, and I learned enough about fishing to contribute to the menu. After

Tom encouraged me to get my Guide's Licence, I was elated to be able to wear the distinctive badge on the side of my cowboy hat.

Meanwhile, Tom got busy setting up something more permanent on the lakeshore than just tents. During the previous winter, using skidoos and towing sleds, he and Bill had hauled in a huge amount of building supplies. skidoos, which had first appeared on the scene in 1959, were still a novelty in 1964. From Tom's stories, it sounded like they'd had enormous fun with them in their pioneering efforts to open up the Rockies in winter. He had a picture of Bill sitting astride a skidoo with a lasso at his side, looking exactly like the cowboy he was. I was sorry to have missed such excitement. But now, I would at least get to work with Tom building the permanent camp. Everything necessary for a small cookhouse and accompanying dining area was already in place. Both the frame tents for guests, each equipped with a pot-bellied stove, and the cookhouse needed to be supported on a base of cut logs on which plywood floors were levelled and nailed. The logs were first cut from trees felled in the bush some way from the camp, then skidded to the lakeshore. This required a choker chain wrapped around one or more logs and joined to a swingle tree (a metal bar used to balance the pull of a draft horse when hauling a load). Two separate chains led from the swingle tree to each side of the horse, where they were attached to an elaborate harness gear that enabled the animal to apply its full strength to pull the load without injuring itself. Only the largest, most powerful horses were chosen for the job.

For me, it was new and exciting work. There was little room to manoeuvre in the thick bush, and at first, I found the commands we used to control the horses confusing. And when I got confused, so did the horse, but Tom soon put me right. I quickly found out that, once moving, it was best to keep going as fast as possible. Most of the effort was spent in getting the logs moving in the first place. At some level, the horses understood that speed increased momentum and made pulling easier. Running along behind, holding the traces and trying to see far enough ahead in the thick bush to steer, as much as possible, an unobstructed route, I had to jump constantly from one side of the logs to the other to avoid them smashing into my legs. I had lots of close calls but thoroughly enjoyed developing the dexterity that the haul-outs demanded.

Building the tent frames and cookhouse with Tom was my first, and last, real experience of carpentry. I got a lot of satisfaction from watching the structures progress under our hands and working with Tom was always great fun. Laughter was the grease. One time, Tom and I were standing on a temporary scaffolding composed of a plank hammered into standing studs. We were leaning over the top of a wall-to-be while hammering on the plywood siding when, without warning, the scaffolding gave way. Luckily, our arms were hooked over the just-constructed roof rafters. The initial shock of falling into space lasted only an instant. Two bodies were now suspended in the air, but safe. A second passed while we looked down at two pairs of legs waving in the void. Tom twisted his head to catch my eye and exclaimed, "Good heavens!" Compared to the super-original streams of cusses I had grown to enjoy, Tom's old-fashioned exclamation was wonderfully unexpected and gut-bustingly funny. I have since spent a lifetime using this totally mild remark, delivered in his exact tone, but only when an event is serious enough to warrant it.

Once the camp was ready, Tom called in Meg, an experienced bush cook and a delightful, unflappable person to have in camp. "She's going to be a heavy water user," he panted as we lugged up innumerable buckets of water from the lake. But we never complained. Meg's food was abundant and excellent. There is something about a breakfast of bacon and eggs served with maple syrup and hot corn bread baked in a wood stove that I doubt I will experience again.

The meals she served to the whole camp when we were in full swing will never be forgotten, but one in particular stands out in my mind, even after all these years.

"We've got Glenn Hall and some friends coming next week," Fay announced one day. There was an excited glint in her eyes while she watched for my reaction.

"Oh yeah," I said. "Who's he?"

She looked at me with a mixture of scorn and amazement. "He's only the most famous goalie in the history of hockey." The name rang a bell, but I had never developed an interest in professional sports. No one in my family was a pro sports fan. But Fay was right. Glenn Hall really was the most famous goalie in history. I confirmed this a year later when I saw him, or rather a facsimile, on display in full goalie gear at Madame Tussaud's Wax Museum in London. I could also confirm that it wasn't a bad likeness.

Glenn's fame certainly preceded his arrival in Jasper. It seemed like the whole town knew he was coming. I learned that he was with the Chicago Blackhawks and had recently set an endurance record of 502 consecutive games that might never be broken. When I met him in 1964, he was thirty-three and at the height of his fame. Checking Wikipedia while writing this in 2020, I see that, in addition to three Stanley Cup wins, his career brought in awards and accolades that would fill pages. But at considerable cost. Protective masks were not mandatory at that time, and Glenn never wore one (except once or twice at the end of his career in 1971). As described by one sportswriter, his face had been "carved like a pumpkin" by stray pucks. Over time, some 250 stitches had been necessary. He clearly understood how dangerous his profession was. His nervousness, it was said, gave him such a pale complexion that his nickname, Mr. Goalie, was sometimes changed to Mr. Ghoulie. I can confirm that was not just a rumour. He was also known to vomit before the start of every game, after which he drank a glass of orange juice.

Magic Travels

Though no hockey fan, when I met him at the corral before we went to the Tonquin, I was awe-struck at being in front of such a superstar. He showed no signs of behaving like one. He was gently diffident and friendly, and to my sixteen-year-old eyes appeared almost elderly. I warmed to him immediately. He was indeed pale, and, at least at first, his bashed-in face and curious teeth were hard not to stare at. I felt shock mixed with pity when, at the fishing camp later on, I noticed how shaky his hands were.

It was my job to take the party up to the camp. We had four saddle horses for Glenn and his friends and three pack horses, two of which carried the men's fishing rods, other gear, and assorted supplies. The third pack horse, Sally, had been carefully chosen for her gentleness and reliability and packed with particular care. Despite soft wrappings in the panniers, the clink of bottles was audible as soon as we started down the trail. Not wanting to take any unnecessary risks, I rode in front of Sally and her precious cargo while leading her by her halter rope. Picking my route carefully, I took special care on the river crossings. In spite of what appeared to be the start of a booze-filled holiday, there was never any drunken or rowdy behaviour from Glenn and his friends.

Meg, an avid hockey fan, was over the moon to have the likes of Glenn Hall dining at her table. One evening, she prepared a superb roast of beef with all the trimmings. After a long day fishing, the meal was much appreciated. Our group huddled around the small table in the cookhouse and treated Meg's bounty as a festive affair. It might have been Meg who brought up the subject of magic and gave me a natural opportunity to perform in the afterglow of the feast. I fetched my case, and with the help of the prevailing mellow mood, had no difficulty making a good show of it. To my delight, Glenn seemed especially pleased and interested. He was greatly impressed when I stole his watch. After I was finished, he slyly asked me for my deck of cards. He wanted to perform a trick himself. First shuffling the deck, he proceeded to deal out three cards face up in a line, continuing to add six more rows to make three separate columns of seven cards each. His friends were laughing but also seemed impressed by his audacity. I knew the trick and felt a surge of embarrassment. It was often the first trick a child was taught, being completely self-working and requiring no skill.

You can try it yourself. Ask someone to pick a card—any card—in one of the three columns. Don't ask what the card is (let's say it's the three of hearts), but do ask which column it's in. Let's suppose it's in the first column. Now, scooping the three columns separately into three piles, you put them all together into a single pile making sure to put the first column between the other two. The twenty-one cards are then dealt once more in exactly the same way as before. Again, you ask which column contains the chosen card. They are then scooped up once more with that column again placed in the middle. The process is then repeated for the third and last time, and you then announce that you will now find the chosen card. Dealing from the top, you turn each card face up sequentially. Secretly, you count them as you place them on the table. The eleventh card will inevitably be the selected card, and you can triumphantly announce that it is the three of hearts.

What was I to do? It was too late to back out. Glenn was already asking me to select a card in one of his columns. I was going to have to pretend that I didn't know the trick. Obviously, it wouldn't do to reveal that I thought the effect scarcely worth performing. With only a slight but involuntary smirk and an inward sigh, I mentally selected the four of spades and waited to be asked which column my card was in.

"And which column is your card in, Patrick?" I pretended to give the matter some deep thought and with an air of reluctance pointed to the one nearest him. He quickly scooped them up, deftly putting the chosen column, as I knew he would, into the middle. Two more times the cards were dealt and the same ritual performed. And now for the grand finale. Glenn started dealing out the cards face up in a pile in front of him. As Glenn no doubt was also doing, I silently counted them as he dealt. Eight, nine, ten, eleven... and down went the four of spades. But wait a moment; Glenn was still dealing them out. My four of spades was already buried. Confused, I kept watching. At card number sixteen, Glenn paused. Fingering it, but not yet turning it over, he said, "I'll bet you my saddle and a pair of gloves that the next card I turn over will be yours." Of course, I knew he was as wrong as could be!

"Okay! Sure, I betcha!" I almost shouted, not knowing if I even understood how such a bet could work. Unhurriedly, Glenn stopped fingering his sixteenth card and proceeded to slide cards away from his discard pile on the table until he came to the four of spades. He picked it up and turned it over.

I had been utterly and royally suckered! As Glenn turned over my chosen card, I realized he knew he had been leading me by the nose. He knew I had totally fallen for it. I'm even sure he knew the scorn I had been feeling when he first started dealing the cards onto the table. The laughter and enjoyment around the table was intense. The bet itself was bogus, but I did tell him I was sorry I hadn't kept his watch. Ever since then, when the occasion seemed right, I have performed the Glenn Hall Sucker Effect with great success and fond memories.

∞

And so the glorious summer slowly passed. Compared to the previous year, no rain ever seemed to fall. One day back at the barn with Tom, I spied a thin, elderly-looking man with his back to us, though elderly, to be fair, is an understatement. Ancient would be more accurate. He was crouched over, apparently examining the lowermost rail of the corral. Tom put his finger to his lips and, creeping up behind, he extended a finger and scratched the hearing aid nestled behind the old fellow's right ear. In surprise, he jumped up and whirled his head around. Tom leapt out of sight to the left. The old man went back to his musings. Again, Tom scratched the hearing aid. Annoyed now, the old fellow leapt up, took the hearing aid off, and examined it carefully. Then, shaking his head in puzzlement, he slapped it several times against the palm of his hand and put it back on.

"Jack!" shouted Tom. Turning around in surprise, Jack focused on us. "Tom! What the hell are you doing here? Goddammit, I never heard a thing. It's this fool hearing aid. Never did work properly. You know as well as I do, I don't need the thing anyways. I just put it on to keep Mary happy. She says I never listen to her." At this, he began to laugh, his toothless mouth emitting a high-pitched chuckle.

"Jack, this is Patrick. Remember his father? You once packed for him," said Tom.

"Who? What?" said Jack.

"Turn up your hearing aid," shouted Tom.

"Okay, okay, I heard you. No need to get ornery. Who'd you say this young feller was?"

Tom repeated the introduction at the top of his lungs.

"Digby's boy?" Jack shouted back. He grabbed my hand and shook it vigorously. "You know, your old man never could get it straight. I kept telling him, but he wouldn't listen. After all them years I've spent with fool geologists, you'd think he'd have more sense. I told him, Diggers, I said, listen to me, and you'll get it right. It's goats on the Devonian and sheep on the Cambrian. And he went and put it the wrong way round on that map he made. Never could figure out what he was up to. I could've showed him where there was gold, but he wouldn't listen. I'm still going to go back there and get it. It's there for the taking, and I know where it is. I could've made your dad rich, but he wouldn't go. I guess I'd better get back there before I'm too old."

This was Tom's uncle, the famous, or infamous, Jack Hargreaves. My father, in his own repertoire of field stories, had often described how his packer and guide from long ago (the very same Jack) took a keen if somewhat zany interest in the work of mapping Rocky Mountain geology. The rock layers (stratigraphic units) were mapped by age, which was determined by the fossils they contained. Cambrian and Devonian were names given to subdivisions of the geological time scale. Jack had listened in on many geological debates among the various scientists in his field party. Geologists are renowned for their enthusiasm, and discussions can quickly become heated. A typical subject was whether a stratigraphic unit was likely to be Cambrian or Devonian in age. Jack seems to have developed a theory of his own that mountain goats preferred newer Devonian rocks and sheep older Cambrian rocks. If true, that would provide one more piece of information making the rocks easier to map. He may have just enjoyed getting in on an argument, but it was possible that, in becoming familiar with what the geologists were doing, he had indeed made such an observation. There are examples of slight changes in rock composition or chemistry that cause burrowing animals to favour one formation or another, and the type of plant assemblages can sometimes differentiate rock types. Good field geologists will use all the observational help available in the construction of their maps.

I had already heard a lot about Jack from Tom, although I had never heard him called anything other than Old Jack. He was certainly getting on, but in his day, Old Jack had been one of the West's foremost packers. On retiring, he had left seven horses to Tom, thereby providing the start for Tom's highly successful guiding and outfitting business. I had spent many glorious evenings, a fire burning cheerfully under the vaulted flap of our lean-to, listening to Tom repeat the extraordinary yarns of his uncle. If only I had written them down! They may be gone forever now.

Tom knew how to tell a good tale. I may not remember the stories, but I remember our laughter at the antics of Old Jack and the various horses, bears, or dudes that he had dealt with in the course of a long and adventurous life. One fact was certain. Old Jack never had any use for doctors or, for that matter, anyone connected with the medical profession. It's more than likely that he was simply frightened. At any rate, his friends had to knock him unconscious before they were able to admit him to hospital when he had appendicitis. The story goes that the doctor was waiting with the anaesthetic at the ready as they hauled him in.

However, maybe Old Jack knew the doctor he was being forced to visit. If it was the same one whose tender mercies I was soon to experience, Jack had every reason to be scared. I was alone with Tom in the Tonquin getting ready for another round of visitors and cutting up logs for firewood with chainsaws. In a moment of carelessness, I allowed the still-spinning chain to graze my upper thigh. It tore through my jeans and made a minor, but messy, gash across the surface layer of skin. I washed and bandaged it at once, reflecting on how lucky I had been. An inch deeper or higher would have been, at the least, life changing. Then, I switched to an axe. Determined to do one last cord before quitting for supper, in yet another careless moment, I swung an ill-aimed blow that glanced off a knot and buried itself into the outer edge of my right foot. Staring in horror at the semi-circular gash in my shoe, blood already seeping through it, I had two thoughts: "Well, that's a couple of toes gone," and "Thank god, I don't have my new cowboy boots on."

Tom reacted quickly and brilliantly. He appeared as if by magic with the first aid kit I had already used for my thigh. Commanding me to lie on the ground, he pulled my foot between his legs with his back blocking my view. For all the world, he might have been shoeing a horse. He made a neat and effective job of bandaging the wound, but by the time we finished supper and got ourselves ready for bed, I was feeling the full effects of what I had done. It was too painful even to rest the weight of my sleeping bag on the bandage. Tom rigged up a small box, which he put inside the bag, to keep anything from touching the foot. That stopped the worst of the pain, but I passed a miserable night unable to sleep.

In the morning, I realized I was going have to get to the hospital in Jasper. We had visitors coming the next day, and there was still a lot to do. I could not ask Tom to leave the work in camp unfinished in order to take me back to town. I assured him I would be all right, and after he saddled a horse and got me onto it, I was ready to go on my own. There was one problem. With one foot bandaged and wrapped in plastic bags, I was unable to use the stirrup on that side. The pain would have been too much anyway, but it made for a long and uncomfortable journey especially when the last part of the trail descended steeply down a series of switchbacks. It wasn't until then that I realized how much more comfortable going downhill was when you took the weight on your legs rather than your crotch.

The trail ended at the road into Jasper. I had been planning to ride all the way, but by now I was in such pain that I was beginning to feel delirious. Continuing on horseback seemed impossible. Dismounting, I hobbled about on one foot until I found a stick long and stout enough to use as support. Then, pulling off the saddle and bridle, I hid them in the brush and let my mount go free to fend for herself. Hopping out onto the road, I waved my stick at the first car that came along. It sped up and disappeared around the next bend. Dimly concluding that maybe the stick looked threatening, I threw it away and managed to balance on one foot until the next car came along. Luckily, it stopped. I have no memory of the ride, but I must have been taken directly to the hospital, dosed there with something strong for the pain, and put to bed. By then, it was pretty late. The doctor would not be available until the next morning.

Although I was in no position to be choosy, I already knew a fair bit about the current doctor in Jasper from things Bill had told me about him. He was a gruff German called Maushacker with few bedside manner skills and, according to Bill, even fewer medical skills. I should add though that, like Old Jack, Bill was not keen on doctors. He had told me the Doc was known to be more interested in horses than people. He had sometimes persuaded Tom to let him take part in castrations and other procedures that occasionally required medical help, but Tom, said Bill, hadn't cared for his methods. Quite apart from the question of Dr. Maushacker's competence even as a horse surgeon, it broke various ethical rules for a medical doctor to work on animals.

The hospital was Roman Catholic and the nurses all nuns. After a drugged sleep, I was wheeled into an operating room early in the morning and left alone on the gurney to await Dr. Maushacker's arrival. A new wing was being added to the hospital, and one whole side of the operating room had been torn down. A temporary wall of clear transparent plastic covered the empty space. Soon I became aware of a group of workmen peering through the plastic. Wearing an all-too-revealing hospital gown, I felt hugely self-conscious and embarrassed, but they only waved cheerfully at me, giving me thumbs-up gestures and other signs of sympathy and encouragement. Perhaps they had already watched the antics of the doctor I was about to experience. But there must have been *some* rules in place to protect patient privacy because they quickly dispersed when the doctor and two white-gowned nurse-nuns entered the room.

The Doc, I am sorry to say, completely lived up to my fearful expectations. He was short and square with large hands and a grouchy expression that never changed. He looked with scorn at the now dishevelled bandaging that Tom had constructed forty hours previously. Without preamble, he yanked it off. As the dried blood was ripped from the wound, I thought I would pass out from the pain, which seemed to set my whole body on fire. He examined the foot, muttering to himself and then announced that, yes, he could probably save both toes. Evidently the axe had passed between the third and fourth toes, cutting skin off their sides as it entered a short way into the foot itself. To save them, he was going to need to take some skin off my left thigh to graft onto the toes. Why my *left* thigh, I wondered, since we were dealing with my right foot. He pulled aside my gown and saw the bandage I had applied after the minor altercation with my chainsaw. This seemed to annoy him greatly. Again he yanked off the bandage. This didn't cause as much pain as before, but it was nothing to sneeze at either.

It seemed clear to him that I was to blame for eliminating my left thigh from consideration. Now he would have to take skin from the right one. He didn't explain why this bothered him so much. Jabbing a syringe into the area, he waited a few minutes, then glowering at me, he prodded the skin.

"You feel that?" he asked.

"Yes," I said.

"That's impossible; it must be frozen by now." Taking what appeared to be a potato peeler, he used it in the same way, removing the top layer of skin in one agonizing, but mercifully quick, stroke. Then, after waving the gruesome piece of bloody flesh hanging from a pair of tweezers in front of my eyes, he popped it into a pan. I had all but passed out by now, though not enough to realize that he had abruptly left the room. I became aware of one of the nurse-nuns bending over me, another syringe at the ready.

"Don't let him know I have given you this," she hissed into my ear.

I felt another jab go into my arm. I have no idea what it was, but I quickly felt an overwhelming sleepiness and a blissful release from pain.

The rest of the operation must have gone well. I was aware of every stitch being put into my toes and foot, but I was calm now and, as they say, feeling no pain. The next couple of days were spent in a ward crowded with injured cowboys from the Jasper rodeo, which was in full swing that week. They kept the hospital staff hopping, and maybe that explained some of Dr. Maushacker's gruffness. If Old Jack and Bill were typical, maybe cowboys and doctors were just not designed to get along.

Old Jack, despite his distaste for doctors and hospitals, actually took the trouble to come and visit me in the ward. After standing in the doorway for some time, staring at me and grinning toothlessly, he advanced cautiously to my bedside. Before I had any idea of what he was up to, he pursed his lips and let fly with a perfectly executed spit directly onto my bandaged foot.

"That'll fix it," he croaked, breaking into wild laughter that turned every head in the ward. "Now you'll be fine. Just make sure you tell Fay that Old Jack came to see you. Tell her Old Jack fixed your foot once and for all. She'll laugh! She'll understand!"

And so she did, both laughed and understood. My foot healed remarkably quickly, and I have never since had a problem with it. I was soon back in the saddle, enjoying my last few weeks of being a cowboy. Those weeks marked the end, I realized as the years passed, of the most formative and marvellous two summers of my life.

CHAPTER 6: EUROPE (1965)

In writing down my life as I remember it, I am forced to conclude that, as far as anything really mattering was concerned, I was a nerd. Here I was, seventeen years old, bang in the middle of the swinging sixties, yet scarcely aware of how much the culture around me was changing. I wore glasses and never could figure out the lyrics to pop music. I was an appalling dancer, couldn't keep a beat, and was shy of girls. I couldn't shake the notion that classical was the only music worth taking seriously. Even though I couldn't help liking a lot of the music most other teens were crazy about, I suspected that really I shouldn't. My brother, Ian, older by four years, was the cool one. He knew how to dress—button-down collars and cuffless trousers—and listened to the hippest music. His girlfriends were all drop-dead gorgeous. I longed to be like him, but it just never seemed to work for me.

It was an exciting time all the same. Pot was beginning to make its entry into the suburban middle classes. That was fun. My friends shared with me a dim realization that education was important and required hard work and that we would all need to be gainfully employed in the not-too-distant future. Still, there seemed to be a more fundamental question: *What on earth should I do with this life?* The Vietnam War was in full swing. Student protests were being brutally quelled in the States. The Cold War was a constant source of anxiety while also providing endless entertainment through the books of John le Carré, Len Deighton, and many others, as well as countless spy movies and TV series. Looming over everything was the prospect of an all-out nuclear war that in all probability would happen accidently. It was clear that adults had caused the problem and that young people ought to be in charge. The likes of Pete Seeger, Bob Dylan, and other protest singers were our Pied Pipers. Wars needed to stop, pollution was beginning to be recognized as a global threat, and even climate change and mass extinctions were being discussed, at least around my parents' table.

Since old people were obviously responsible for the world's ills, enlightened youth were the key to its salvation. But what to do? One overriding principle seemed paramount: choose a path through life doing something you loved. To pursue money and security was tantamount to submitting to the forces of evil. The worlds of business and commerce, even law, were sell-outs to the other side, which we despised. There was a profound belief that working at what you liked would ensure you would be good at it. The money would take care of itself.

In the meantime, the choices could wait. There was time to travel and see the world. And thanks to ever-cheaper airfares, seeing the world was something almost any young person could now do. Explore the possibilities and have fun while you did it. I didn't have any problems with that philosophy. The most popular student destination at that time was Europe. It was known to be cheap. Bookshops were full of titles like *Europe on Five Dollars a Day*, unlimited Eurail passes were an amazing bargain, and above all, my friends and I were Canadian. Conventional wisdom held that American students were already flooding Europe, and their wealth and arrogance were not appreciated. The standard uniform of the student traveller included a small flag of your country sewn onto your packsack. Word had it that Americans removed theirs when they realized that it ensured a less-than-friendly welcome. It was even rumoured that good money could be made selling Canadian patches to American travellers. It was highly desirable not to be taken for an American, while Canadians apparently were received warmly throughout Europe.

Experience showed me there was an element of truth in these travellers' tales. Having decided to continue with hitchhiking as my preferred mode of travel, I noted the nationality of the licence plates on the cars that stopped for me. Dutch, French, Belgians, and even British drivers all made me feel exceptionally welcome the moment I identified myself as Canadian. I soon figured that this could best be explained as an aftermath of the Second World War. For a seventeen-year-old from Ottawa, the war was a long time ago, but in Europe twenty years had not been enough to erase its horrors, and there was a consensus that Canadian heroism had been exceptional right from the start. In contrast, the Americans, as several drivers explained to me, had arrived late and then exasperated their allies by taking the credit for winning it. German drivers were even friendlier, displaying a hospitality that could even seem over-the-top. I decided they must have been eager to demonstrate post-war goodwill toward former enemies and show that not all Germans were bad guys. They certainly convinced me of *that*.

Another attractive element of such travel was the opportunity it gave to imagine oneself as the carefree, but world-weary, wandering troubadour romanticized in so many of the folk songs we listened to. As Peter, Paul, and Mary sang: "Why don't you help me brother? I'm a stranger in your town. Why don't you help me sister? And maybe

I'll settle down." A guitar across a traveller's back, ideally with the ability to play, was a common sight on the road. It was rare to stay at any youth hostel without a smoke-filled hootenanny starting up in a dorm or stairwell. I had no skills as a musician. What I did have was my magic case. This was as easily strapped onto the packsack as a guitar. It would be an alternative way of meeting people and would help me develop as a performer at the same time. As I wrote to my parents from Berlin (July 17, 1965), "Thank god, I brought my magic. Although I haven't made any money, it has enabled me to make many friends. At every youth hostel, I have done a show. Without it, I would be very lonesome."

Plans for the great adventure took shape throughout my final year (Grade 13 in those days) at Glebe. With Guy Sprung, who was keen to follow suit, I could not have had a better companion. Although I had taken German and French in high school, I was a long way from speaking either with any fluency, but Guy spoke both equally well. We would also be able to stay for a few days with relatives of his, one family who had a farm near Hannover and another who lived in West Berlin. These visits were sure to improve our time in Germany in countless ways. I had some family links too, thanks to my father's geological connections. The most important was Dr. Paul Sartenaer in Brussels, whom I already knew from his many visits to our home in Ottawa. A distinguished paleontologist at the Royal Belgian Institute of Natural Sciences, Paul, like my father, was a specialist in brachiopods, a shellfish group whose fossil remains were (and still are) important in the search for oil. With his pleasing Belgian accent, dapper appearance, and precise way of talking, Paul always made me think of Agatha Christie's Hercule Poirot

Though perhaps not as fat as that famous sleuth, Paul was a renowned gourmet. Legend had it that he could tell you whether a particular wine came from grapes growing on a north- or south-facing slope. He frequently boasted that Belgian food was the best in the world, combining French quality with German quantity. At one point, Paul decided that examining present-day coral reefs would help him understand what he was seeing in the rock record. Rather than take any old dive course, he went straight to Jacques Cousteau at that famous diving pioneer's headquarters in Monaco. The two got along famously. As soon as Paul had mastered diving, he decided to tackle the Mozambique Channel between Madagascar and Africa, known to harbour the largest number of the most dangerous shark species in the world. Strangely, he was unable to find a companion to dive with him. Unperturbed, he dived alone.

"But Paul," my father once asked him at the dinner table, "what did you do about the sharks?"

"Ah!" said Paul, "Zay are everywhere," waving his arms with enthusiasm, "but zay are zo beautiful! Zo lovely to watch! Of course, I had my work and couldn't spend ze time to pay them attention. But always I carry a short, strong steek stuck in my belt. Sometimes, when zay came a leetle bit close, I had my steek all ready. Then, I would beef zem sharply on ze nose. Zay deed not like zat! Zay would leave me alone after zat, and I could get back to work."

My brother was another source of useful connections. He had spent a summer working in Wood Buffalo Park in northeast Alberta, building corrals and having various adventures with the buffalo that were put into them. Later, as a hotshot skier, he wangled a job with a French company (CIM) teaching skiing at Val d'Isère. There, his buffalo stories evidently became famous. He was featured in tourist magazine advertisements, inviting future guests for après-ski relaxation with Ian McLaren and his stories of wild escapades in the Canadian North. Through Ian's connections, we now had a place to stay at the CIM clubs in Paris and at Val d'Isère.

Guy and I flew to London first and immediately bought tickets for the boat train to Paris. To kill time before departure, we spent a few jet-lagged hours listening to the speakers at Hyde Park Corner. We had been intrigued by this curious wonder of British life where anyone could rant about anything at all without fear of censorship. We knew that many of the speakers were fixtures and could be counted on to be there, raving on about their favourite subjects. The speaker we spent most of the time listening to was elderly, large, muscular, totally bald, and heavily tattooed. He didn't seem to like anyone much and spent most of his time choosing random nationalities and screaming insults about them. The large crowds he attracted ensured that on any given day some members of his audience would probably belong to the nationality he had selected to abuse.

Some reacted angrily, but most just smiled. The less effective he was in getting a rise out of his victims, the angrier and more abusive he became. On that day, it was only when he turned on some Americans that he got the reaction he desired. "You Americans are TRIPE!" he blared. "You come over to our country, swaggering about, thinking you know everything, waving your precious money in our faces. What good are you? Why don't you stay back in that hellhole of a country where you belong? YOU ARE NOT WELCOME HERE!" To his evident delight, the crowd contained quite a large group of Americans that afternoon. It was fascinating, if not educational, to listen to their vehement anger directed back (to no avail) at the speaker. They were, after all, the only nationality that had responded in any way, other than by rueful laughter or even enjoyment, at the antics of an entertainer, albeit mad as a hatter, who had finally achieved the reaction he was seeking.

The 10:00 p.m. boat train took us to the Channel ferry. Unrolling our sleeping bags on the deck, we caught a few hours' sleep before boarding the train from Calais to Paris. There we made our way to Avenue de Wagram, one of the thirteen streets and boulevards

Magic Travels

entering what is justly known as the craziest roundabout in Europe, centred around the Arc de Triomphe. To watch from the top of the monument the chaotic and ever-changing patterns of cars jockeying for position is a mesmerizing and entertaining experience that I would recommend to anyone. After only a few bum steers, we found a little empty night club called Le Cave that belonged to CIM. A concierge fetched Monsieur Constant, my brother's connection, who turned out to be as friendly and welcoming as promised. He even scrounged up a couple of camp beds to spare us from another night on hard floors. The club was closed for the season but, giving us a key, he invited us to use whatever facilities we needed and to come and go as we pleased.

And what did Guy and I do in Paris? What I can only suppose any two young men free from school for the first time, with no further parental constraints and acutely aware that they were almost grown up, would do. Our first supper in our own nightclub consisted of ham, cheese, a bottle of wine, and a baguette that tasted better than any bread we had eaten before. The wine was cheap and raw, but we pronounced it excellent. We had the same again for breakfast. We roamed the streets, explored the usual tourist traps, discovered nightclubs in the sleazy Pigalle district, sat in cafés, and revelled in the atmosphere of the Latin Quarter and Montparnasse.

Throughout my teens, I had been an avid reader of George Simenon's Inspector Maigret novels, so I felt I already had a good grasp of what Paris was like. I wasn't disappointed. Simenon's detective stories seemed to describe exactly the Paris I was seeing. I loved the city. In one of the novels, Maigret took to drinking Pernod for the duration of the case he was working on. Having no idea what Pernod was like, but feeling very Parisian and sophisticated, I ordered some. A small glass containing yellowish fluid was set in front of me. Wasting no time, I took a deep swig and nearly spat it out. It was horrible. The waiter, with an expression of supreme contempt, produced the carafe of water that needs to be added to it. I did so and was able to enjoy a milky, pleasant drink tasting of liquorice. That was the first of my many encounters, as a naive foreigner, with the famously haughty manners of French waiters in bars and restaurants.

With the first few joyous Parisian days of our European adventure finished, we split up to hitchhike to our next destination, Rue des Egyptiens in Brussels, where Paul Sartenaer had his flat. Taking the metro to the end of the line, I picked up lifts quickly and made my way northward through Beauvais and Abbeville, arriving at Montreuil to spend my first night at a youth hostel housed inside a castle. I spent the evening enjoyably with a fellow Canadian at an open-air circus in the town square. The next day, I arrived at Rue des Egyptiens in the mid-afternoon to find Guy already there. Paul was wonderfully affable. Taking us to dinner at one of his favourite restaurants, he explained that he had moved in with his brother to give us the run of his flat. He also gave us the keys to his Volkswagen and insisted we use the car to visit Bruges and Ghent. We were overwhelmed by his generosity but, as events transpired, were only partly able to follow his advice on what to do and see.

∞

Paul's flat was surprisingly small. It was also quite spartan in its furnishings, though every piece was excellent and had clearly been chosen with great care. On the first day of being left to ourselves, I called Guy over to a kitchen cabinet where I had discovered a pair of cut-glass wine goblets.

"Take a look at these Guy. Have you ever seen glasses as beautiful as these? They must be worth a fortune!"

Paul was what, in those days, was known as a confirmed bachelor, and I knew he had various mistresses, one of whom was a Russian ballerina. It was easy to imagine these two exquisite glasses as essential equipment for intimate tête-à-têtes when some dazzling catch was being irresistibly seduced, overwhelmed by the magnificence of her goblet (and, presumably, its contents). Guy was as impressed as I was.

"Listen," he said. "Whatever happens, we must not use these. I'll put them at the very back of the cupboard out of harm's way." Which he did.

I'm sure that what happened next can only have been Guy's fault. He must surely have placed the glasses in a dangerous position. Otherwise, I cannot explain how, on pulling something else out of the cupboard, my sleeve caught on one of the goblets and pulled it from its hiding place. I watched in dismay as it seemed to fall in slow motion toward the stone-tiled kitchen floor where, with an almost musical tinkling like chimes, it smashed into microscopic pieces. The chimes brought Guy running into the kitchen, and we both stared speechless at the wreckage. We were mortified. Here we were, guests of a generous and dear friend of my parents, given free rein to stay in his apartment while we toured the wonders of Belgium in his car, and how had we repaid his kindness? We had broken one of a pair of beautiful and perhaps priceless wineglasses, leaving its mate stranded and essentially useless. That's what we had done.

The glass had to be replaced. That was obvious. We were leaving in two days, and Paul was not coming back before then. We had already said our goodbyes and arranged to leave the keys to the flat and car with the concierge. Instead of being idle tourists in Belgium, we now had an urgent mission. Packing the intact glass in a shoebox lined with cotton wool and with the help of the concierge, we made a long list of potential suppliers of such glasses in Brussels and its surroundings. A lot of time was spent on the phone to determine the likelihood of finding such a glass. The ferocious traffic and our inexperience with Belgian driving regulations, either official or merely understood by common knowledge, were a challenge.

A narrow escape occurred when I failed to recognize the European *Do Not Enter* symbol across what I thought was the entrance to an underground parking garage. It was for pedestrians only. With the brightness of the day outside, it was like driving into a black hole. That explains why I didn't see the flight of stairs leading down into the depths of the parkade. A subconscious warning system must have kicked in, or perhaps I caught a glimmer of the metallic strip on the top stair tread. In any case, something made me slam on the brakes, and I stopped within a millimetre of going over the top step. Guy, sitting beside me, produced a torrent of loud and colourful expletives. I still shudder when I imagine Paul's Volkswagen hurtling uncontrollably down a flight of stairs. We might even have broken the other glass.

Over the next two days, we must have visited every store in Brussels that sold fine crystal. At each one, we got another lead on where such a glass might be found and walked out of every one of these empty-handed. It increasingly seemed like an exercise in futility. At the end of the second day, we dragged ourselves into the last shop on our list, our final chance for success. We needed to leave Brussels first thing the next morning. This shop was probably the most upmarket of all those we had tried. The owner was wonderfully sympathetic to our plight. Then he took one look at our specimen and announced that this type of glass had not been made for at least ten years. He stated categorically that we would never be able to find another like it. But he did have a suggestion. From a display case of extraordinarily beautiful goblets, he picked out one that, he explained, complemented ours in such a way that the two halves made an even better whole. He set them side by side. The pairing was perfect, and we felt sure that Paul would agree. Unfortunately, however, the owner couldn't possibly sell his display piece. He would have to order another, which would take a week or two to arrive.

Magic Travels

The man clearly loved his business, and we trusted him implicitly. We paid for the chosen mate (very expensive!) and asked him to keep the original with him. We would ask Paul to collect both when he returned to Brussels. That way, just in case he wanted to choose another, he would be free to do so. We shook hands all round and returned to Paul's flat to write him a letter, which we left with the concierge, apologizing for the accident and providing detailed instructions on how to retrieve the two glasses.

Some years later, Paul was in Ottawa visiting my parents again. I was invited to dinner. With some trepidation, I brought up the subject of the broken goblet. "Oh yes," said Paul. "I remember the concierge giving me your letter. But I am zo sorry. I forget all about it. I originally had a set of eight, but you know what things are like. Zay break zo easily! I break one, then two, the cleaning woman breaks another. Just a little tap and that was enough. As ze years passed, all I had left were two. Not enough for a dinner party, zo I hid them in ze cupboard and no longer used them."

Our next stop was Helmstedt, right up against the border with East Germany, where Guy's relatives had a farm. Located in the northeast of West Germany, it was 568 kilometres by the shortest route from Brussels. We decided to separate and meet there in five days. I headed southeast into Luxembourg. On my first night, to save money (after the unexpected expense of the goblet), I chose a deserted field and bedded down for the night with my groundsheet wrapped around my sleeping bag. This was not enough to keep me dry in the downpour that woke me in the small hours of the morning. Completely soaked, I stumbled across the field, skidding on innumerable giant slugs that now covered the field and took shelter in a small garden shed that was stuffed with tools and fertilizers. It was not comfortable. At first light, shivering in the damp, I made my way across the border into Germany where I found a youth hostel in the tiny village of Saarburg. There, I was able to dry my belongings and recover at leisure.

That evening, the hostel filled up with boys from a German private school. Not understanding the dining arrangements, I joined their line to receive plates of food being handed out through a sliding hatch in the wall of the dining room. The atmosphere was not warm. The students pointedly ignored me, and I was beginning to feel distinctly uncomfortable and out of place when, without warning, an older man, who I later learned was the headmaster, challenged me. I struggled to follow his rapid-fire German but eventually gathered that he wanted to know what I was doing there and why. This was a private group, and I was not entitled to eat my meal at this hostel. He stopped and glared at me, awaiting an answer.

The silence was absolute. The students stared at me, probably enjoying the confrontation. Somehow, in my halting high-school German, I managed to explain who I was and placate him by apologizing for not understanding how the hostel worked. At last, he grudgingly nodded toward the hatch and turned his back on me. Taking my plate of food, I went to sit at an empty table, but a student caught my eye and motioned me to join him and his fellow students. The ice was broken. The headmaster had gone off to eat somewhere better, and it was soon clear to me that the students despised him. After dinner, they were all going out to drink beer in a nearby pub. Would I like to come?

At the *bierkeller* we quickly located, they took full advantage of their freedom, and the beer literally flowed. In return for their hospitality, I brought my magic case and gave them a show to great effect. Their cheers and laughter were as abundant as the beer, and we became fast friends. It was a wonderfully upbeat way to end a miserable twenty-four hours. Piling out of the pub at closing time,

Magic Travels

I retain a vivid memory of mass urination into the ditch beside a deserted road as we noisily stumbled our way homewards.

After this, I was determined to put my high-school German to good use for hitchhiking. My stock phrases were enough to communicate where I was going and where I came from, but I soon learned to take part in rudimentary niceties with drivers and other passengers who might be in the car. I got used to both German friendliness and fast German driving. I remember one driver who noticed my white-knuckled fist clinging onto the grab handle above the door and announced in careful English, "Do not be afraid!" as he sped past yet another car going only slightly slower down a steep narrow road while rounding a blind curve.

That style of driving at least ensured that in a few hours I had made it all the way to the Rhine, but after a night spent at picture-postcard Oberwesel on the steep side of the river, I went through a bad patch with lifts. By noon, I hadn't made more than a few kilometres when it began to rain. The downpour seemed to change my luck. Before I was even wet, a driver who lived at Koblenz further down the Rhine rescued me. This was a godsend. I needed to pass through Koblenz to reach Limburg on my route to the northeast. What followed was more of the extraordinary hospitality I was getting accustomed to in Germany. Why don't I come to his house for Sunday lunch? I could hardly say no. And besides, I had eaten nothing since the night before.

His wife and five children were, if anything, even friendlier. I was treated like a long-lost friend over a massive meal of roast pork, cauliflower, and sautéed potatoes. An after-dinner magic show was a perfect means of expressing my gratitude. The whole family then piled into the car, and we drove all the way to Limburg. My luck had decidedly changed. I made another three hundred kilometres to Bad Wildungen halfway from Koblenz to Hannover and arrived at the hostel just in time for supper.

Walking the last kilometre down a hot, dusty country lane, I reached the farmhouse of Guy's relatives three days later. Situated outside a tiny village between Braunschweig and Helmstedt, its surroundings appeared dull and uninteresting compared to the scenic beauty I had enjoyed for most of my trip from Brussels. Guy had arrived a few hours before me, and he set about introducing me to what struck me as an unusually large number of sturdy middle-aged women for one household. Probably they were only three or four, but they filled the space with enthusiastic excitement and made me feel extraordinarily welcome. I never really did understand their relationships with each other or where Guy actually fitted into the family tree. It certainly seemed a complex dynasty. The roost was ruled, nominally at least, by a grizzled, cantankerous-looking old man. In fact, he was known as the Old Man. His left arm was missing, lost during the war. Later, watching him at work, I was impressed by how effectively he used a sling that supported a pitchfork while levering the fork with his remaining arm to toss hay or manure. That explained the enormous strength that was clearly visible in the bulging muscles I saw when his sleeve was rolled up. I was assured that ordinary two-armed men were unable to keep up with him.

We had evidently arrived at a unique time in the life of a working farm in that part of the country. One harvest had already taken place, and it was still a little early to start planting a new set of crops. This meant that full attention could be given to Guy and his friend for the entire length of our stay. When I arrived, a pig had already been slaughtered in our honour. The farm was ancient—I don't know how many hundreds of years old—but it was easy to imagine that it could have been there from medieval times. It was also completely self-sufficient. The family ate what it produced year-round. With no pressing work to do for a few weeks, they treated Guy and me with a hospitality that actually became a little oppressive. There were four meals a day. The first three (breakfast, lunch, and tea) were all the same: cold cuts of at least a dozen types of ham, salamis, and pâtés accompanied by thick slices of black bread and various salads, all washed down with beer (well, maybe it was only tea at breakfast). The fourth, the evening banquet, was the proper hot meal of the day, usually a roast with abundant vegetables followed by a dessert that had usually taken most of the day to make.

Our days at the farm were perhaps the only time in my life when I couldn't keep up with the eating going on around me. Teenage boys are meant to be bottomless pits, and I had thought that I more or less fitted that description. Not on this occasion. I began to dread the next meal. The evening feasts felt like some form of slow torture at the end of which the victim was meant to explode. The pressure from both within and without was relentless. Even when my stomach had reached its limit long before, any hint of reluctance to stuff down a second, third or, yes, a fourth helping was cheerfully ignored. The barrage of generosity and goodwill was overpowering, the signs of disappointment on the women's beaming faces too painful to bear. I could not be that rude to hosts who had taken so much trouble to make me feel so welcome.

Two nights before we were due to leave, the women of the family had to attend a village meeting. After the evening meal, bedecked in huge flowery skirts and blouses and trailing the scents of many perfumes, also flowery, they said their goodbyes and departed. The door had no sooner closed when the Old Man immediately sat us down with full flagons of beer and an unopened bottle of schnapps. It was clearly his night out at home. With two, well, nearly men, to enjoy it with. The absence of constant clucking and fussing from the females who were obviously the real rulers of the roost must have been the driving force behind the unexpected ebullience he now displayed. Ordinarily taciturn, he grinned broadly as, with practised movements, he single-handedly opened the schnapps, poured out three shot glasses, and began a rapid, non-stop discourse that lasted the whole evening. I could barely follow, but I understood the etiquette of the occasion. This had one purpose, to get as thoroughly plastered as time would allow. You never knew when a village meeting might be over. The routine was to down the shot glass in one gulp, drink a flagon of beer as a chaser, and repeat. The beer was not for sipping. The next shot of schnapps was already waiting. I wish I could give a coherent report of how that evening turned out. There was a great deal of raucous laughter. That much I know. It may even be that my German rose to the occasion. I vaguely remember feeling remarkably fluent. I do know that the bottle of schnapps was emptied. The beer remained bottomless. I have no memory of the women's return or how they dealt with us.

Next morning, the inevitable breakfast charcuterie was all laid out, and the eating for the day began. Even with the most wretched hangover I had ever experienced, it was still impossible to enact my fantasy of running away from the table and hiding under my bed until my guts and head recovered. Instead, because we were planning to leave the next morning, it seemed important that Guy and I must have a torte for dessert to celebrate our last supper. Immediately after breakfast, the four women disappeared into the kitchen to spend all day constructing what must have surely been the most complex and, if I had been even remotely hungry when it was wheeled out after dinner, the most delectable cake on the planet.[1] The torte produced that day for our supper was as authentic as they come.

But first, there were two more of the standard cold meals to get through, plus a huge hot stew again prepared especially for our last supper. Finally came the torte. Already overfull once again, and still suffering from the night before, I watched with dismay as the vast

---

1   An authentic German torte is described in cookbooks as "a rich, usually multilayered cake that is filled with whipped cream, buttercreams, mousses, jams, or fruits. Ordinarily, the cooled torte is glazed and garnished."

creation was brought in to great fanfare and excitement. No wonder it had taken all day to make. There must have been at least eight layers and between each was a whipped-cream filling with a unique flavour based on different berries and liqueurs. It was magnificent.

But I knew I was facing the proverbial straw that broke the camel's back. I mumbled my admiration and enthusiasm as an enormous first slice was placed in front of me. Excellent though it was, I had finally reached my limit. I couldn't possibly do it justice. After a few mouthfuls and some appreciative noises, I put my skills of misdirection to the best possible use, namely to save my life. By conversing animatedly with Guy, I used his responses to encourage the table to focus on him each time I made my move. At the precise moment when all eyes were directed toward Guy, I surreptitiously spilled a forkful of cake into the napkin on my lap, which in turn was hidden by the tablecloth. As soon as I had dumped the last bit, I excused myself and hurried off to the bathroom, concealing the full napkin behind my body. With a profound mixture of relief and guilt, I flushed it all down the toilet.

I returned to a table of innocent, beaming faces. No one had seen me. Even Guy was unaware of my antics until I told him afterwards. But my overstuffed body got no respite in the morning, our day of departure to Berlin. After one final gargantuan breakfast, I was seriously worried about how long I would be able remain away from a bathroom, especially as we now needed to hitchhike across the East German border. After tearful farewells, we were driven the few kilometres to the border crossing and let out at the side of the autobahn within walking distance of the guardhouse. We were at Checkpoint Alpha, the largest of the crossing complexes along the border dividing capitalist West Germany from communist East Germany. It was close to the height of the Cold War, only two years since the Bay of Pigs crisis in 1963 had brought the world as close to nuclear war as it had ever been. Tensions still ran high.

Checkpoint Alpha was probably the single most sinister symbol of the Cold War era. For us, at least, it epitomized the horrors of the Iron Curtain that divided Germany from the end of the war until the fall of the Berlin Wall in 1989. As tourists heading for West Berlin, we had to take the autobahn corridor through East Germany directly to Checkpoint Bravo at the entrance to the divided city. There we would once again cross the border and enter West Berlin, a walled-in island of freedom surrounded by totalitarian East Germany. The autobahn corridor along which we had to hitchhike ran for 190 kilometres between the two checkpoints.

Guy and I understood that the one thing we must not do was mess around in any way whatsoever with the guards, police, or any other officials that might be responsible for allowing access to the corridor. Answers to all questions must be simple, polite, and to the point. Trying to be smart or sounding the least bit sarcastic could have immediate and severe consequences. All the same, I now had more pressing concerns that had nothing to do with international relations. There, at the side of a busy highway and before proceeding to the checkpoint buildings, the consequences of the enormous quantities of food and drink I had consumed over the last few days came to startling fruition.

Undeniable, unrelenting signals began to emanate from my bowels. Immediate relief was imminent whether I liked it or not. Shouting a few garbled words of explanation to Guy, I sped up an embankment into the welcoming privacy of the woods that lined the autobahn. With my trousers around my ankles, just as I was achieving the relief I sought, two savage German Shepherds abruptly interrupted my labours. In my rush into the bushes, I had failed to notice the steel meshed fence topped with barbed wire that wound through the forest. Now, a pair of well-trained guard dogs were barking their heads off a few inches away as they did their best to rip apart the fence and then presumably me.

Their barking and snarling were soon followed by the sound of running boots as a heavily armed soldier appeared from the trees waving a machine gun and shouting, "Achtung! Achtung! Was machst du hier? Sofort gehen!" (Attention! Attention! What are you doing here? Leave at once!) I did not need a second warning. Stammering useless apologies and pulling myself together as best I could, I fled back down the embankment, expecting a hail of bullets at any moment. I had just had my first encounter with an East German border guard. The sudden terror might even have helped to solve my gut problem.

Mortified and still trembling, I rejoined an anxious Guy, and we approached the checkpoint. The idea was to walk across and start hitchhiking again on the other side. I felt convinced we were being watched every step of the way. I felt certain my altercation with the guard would already be known, with consequences I could only imagine. I left it to Guy to explain things to the guard who examined our passports. With a huge grin exposing broken and nicotine-stained teeth, he winked at us and motioned to a car that was just pulling away. Instructing the driver to stop and roll down his window, he roared, "Nehmen Sie diese beiden jungen Männer mit nach Berlin. Sie haben hinten Platz" (Take these two young men to Berlin with you. You have room in the back.) Then, like a hotel doorman, he held the back door open for us. And so went my second encounter with an East German border guard. We arrived at Guy's grandmother's apartment in no time.

Like the knowledge of Paris I had gleaned from Simenon and Maigret, I had the same sense of already knowing Berlin, but now through the eyes of Harry Palmer (brilliantly played in the movies by Michael Caine) in Len Deighton's Cold War spy books. The bombed-out shell of the Kaiser Wilhelm Memorial Church, left unrestored in the centre of West Berlin as a monument to the horrors of war, spoke of the immediacy of the conflict, yet the city bustled with a hedonistic energy where anything could go and probably did. The underlying current of danger that I also sensed came, perhaps, from the realization that East Germany, the total opposite of the carefree opulence of West Berlin and the almost hysterical hustle and bustle of its inhabitants, completely surrounded the city and was held back only by a rather primitive wall. A heavily guarded wall to be sure, but its presence was a constant reminder of the extreme nature of the Cold War. Guy and I took the permitted one-day excursion to the other side through the iconic Checkpoint Charlie. Nobody could describe the experience better than Jan Morris: "Travelling from west to east through the inner German border was like entering a drab and disturbing dream, peopled by all the ogres of totalitarianism, a half-lit world of shabby resentments, where anything could be done to you without anybody ever hearing of it, and your every step was dogged by watchful eyes and mechanisms."[2]

The bullet pockmarks in the cement walls of cheerless, crumbling buildings and the heavy, wearisome walk of drably dressed inhabitants whose eyes never met your own made our day trip feel like a visit to purgatory. Getting back to the other side was a relief, but somehow its glitter now seemed tainted and almost incomprehensible. Events seemed to conspire with our new mood of disillusionment. It was time to leave Berlin for Munich, but our attempts were unaccountably difficult. The first border guard chased us away with no explanation, and as we trudged back down the autobahn, we were rudely accosted by the West German police. We decided to lay low for a few hours before trying again. A new shift of guards let us through without any problems, and we snagged a lift right away that took us all the way to a small village north of Frankfurt.

---

[2] Jan Morris, *Fifty years of Europe: An Album* (Random House Publishing Group), 1997.

There, we again misjudged the weather by bedding down in a field for what turned into a rainy night. After packing up our sodden belongings in the morning, we decided it would be easier to split up and continue separately.

There followed a week or two of many long and eventful days, sometimes together, sometimes separate. We discovered beer drinking at the Hofbräuhaus in Munich. We were invited to a lodge in the beautiful Bavarian Alps where we attended an outdoor wedding that could have been filmed for *The Sound of Music*. We hiked in the mountains of the Austrian Kaisergebirge and in the even higher mountains of Tyrol, staying in mountain huts along the route. We got lost as we struggled through waist-deep snow while crossing a mountain pass in a blizzard and were stuck in a hut for two days, waiting for the snow to melt enough to follow the trail markers. All the while, I carried my magic case and gave impromptu performances in countless places and circumstances. I became skilled at assessing my audiences and picking the members to engage with and the ones to avoid. I learned the types of personalities that simply did not like the idea that they could be tricked, as if the point of magic was to make a fool of them. Such people could not be trusted. They were the ones who might grab at the props, hoping to reveal the trick (which of course they did). I learned to judge on the fly whether it was best to ignore them or use them as part of the entertainment.

En route to Geneva, on my own again, I had been having great difficulty getting lifts. To this day, I confess to reservations about the Swiss after hitchhiking the entire length of that country without being picked up by one local citizen. French, Germans, occasionally Austrians, even the odd Brit—but never a Swiss! My theory, difficult to prove, was that it had nothing to do with lack of compassion or generosity. It was their obsession with being on time. Clocks littered the towns and villages. Trains and buses left and arrived to the second. Civil unrest occurred if it were otherwise. Stopping for a hitchhiker would simply take too much time. And no Swiss could afford the stigma of being late.

One lift on that leg of my journey led to some memorable consequences. A new-looking Mercedes, the type of vehicle that did not normally stop for a scruffy hitchhiker, unexpectedly pulled over. Even more unexpected was the attractive girl at the wheel. Yes, she was going into downtown Geneva and would be pleased to take me. We were soon Patrick and Jen. She was British but lived with her ex-pat parents in the city. After a pleasant chat and before dropping me off at my hostel, she invited me for lunch the next day at her home. I explained that I was en route to Val d'Isère and would need to get away by mid-afternoon. Would that be okay? Absolutely no problem. She wrote down the address and drove off.

What I learned only when I got there, dressed in cords and flannel shirt and with my by-now well-worn packsack in tow, was that the address Jen had given me was the British Embassy. She was the ambassador's daughter. A tuxedoed manservant opened the door and seemed to have been expecting me, but his quick, silent appraisal was not friendly. I knew the instant I stepped in the door that this was a mistake. Jen seemed untroubled, however, and began introducing me to her parents and at least a dozen middle-aged and elderly guests as if I were to the manner born. To say that I felt out of place would be a colossal understatement. Everything about me was wrong—my accent, my clothing, my age, and how I had met Jen in the first place. I was obviously (I now had a beard) one of those wild youths that fell under the all-encompassing label *hippie*. They were all far too well-bred to advertise their discomfort, but I knew how to read a crowd.

Still, champagne was being served, and I was learning never to turn down freebies. Buoyed by Jen's friendliness and a few glasses of the bubbly, I began to relax. An impressive luncheon was served: vichyssoise, beef bourguignon, salad, and assorted pastries with cream, all washed down with different wines for each course. But I had made another mistake. While riding with Jen the day before, I had entertained her with my funniest stories about doing magic on my travels, even pointing out the case strapped to the back of my pack. With no warning to me, Jen announced during the coffee service, "Patrick is a magician and performs all over Europe. Maybe he could be persuaded to do a show."

I could surmise from the stony faces that greeted this suggestion that the room wanted nothing to do with me, let alone to be tortured with an amateur magic show. But Jen, as the ambassador's daughter, evidently swung considerable weight. There was little the captive audience or, for that matter, the cornered magician could do to get out of the fix we all seemed to be in. The room was quickly rearranged, chairs were set out, and I laid out my equipment on a card table in the front. Several of the guests were already sitting back in their chairs smoking, so I didn't hesitate to open the show by producing lit cigarettes from thin air and taking a puff on the last one before disappearing it into my fist. As I waited for the applause, a stout sour-faced woman in the front row had a sudden fit of coughing. "Didn't anyone tell you I'm allergic to smoke?" she gasped. "It's outrageous to light cigarettes with no consideration for others. You should be ashamed. If you light any more cigarettes, I shall have to leave immediately." I was dumbfounded. No one in the room was putting out *their* cigarettes, but I avoided pointing this out, stammered an apology, and proceeded to my thimble sleights, followed by the three-rope trick. My nemesis was not finished with me. "Oooh," she exclaimed, "I've seen that one on the telly. It was done ever so much better than that. I think he has a separate set of trick ropes. It's all completely fake, you know. I can't think why anyone could actually be taken in."

She was exactly the kind of audience member that can and will deliberately ruin any attempt at a successful show. I was done for.

Yet—maybe it was the effect of several glasses of wine—I didn't care. With a wonderful sense of recklessness and feeling a surge of anger, I pretended to study the audience carefully and announced that my next trick required a volunteer. Before Jen could raise her hand, my eyes fell as if by accident on my enemy, and in my suavest tones, I asked her if she would be so kind. I made it clear that she was uniquely suited for the next effect, and she grudgingly took her place beside me as instructed. Taking two large silk scarves from my case, I shook them out, displaying both sides to exhibit their innocence. Holding each of them by a corner, I tied them together while explaining that I would now pull the scarves right through her body. Then, with just a touch of rudeness, I yanked at the neckline of her blouse and quickly stuffed the knot down her front to be held there between her copious breasts. The ends of the two scarves now dangled from the top of the blouse on either side of her body. Ignoring the woman's evident discomfiture, I asked Jen if she could also come up to help me. She rushed eagerly to the front, and I instructed her to hold the end of one of the scarves while I took hold of the other.

"Now," I told Jen, "when I count to three, I want you to pull hard on your scarf while I do the same in order to pull them right through—sorry, what did you say your name was? Ah yes, to pull them right through Mrs. Beamish. If you'll just stand still, Mrs. Beamish, we are now going to do the impossible and pull the two scarves right through your body. No, no, I promise you, it won't hurt. Are you ready? One... two... three... PULL!"

Perhaps you've already guessed the outcome. Instead of seeing the scarves going right through Mrs. Beamish's body, the audience was now looking at Jen and me holding the two ends of the scarves with a grotesquely large, bright pink brassiere neatly tied in place between them. A weak trick when all said and done, in fact hardly a trick at all since it is patently obvious that the bra had come from inside one of the scarves. Nevertheless, with the right audience at the right time, it can be counted on to yield the intended laughs. This, however, as I well knew, was definitely the wrong audience at the wrong time. You could have heard a pin drop. Not even a snicker escaped from anyone except for the loud giggle from Jen. That was the end of the show. I wasn't even shown to the door. But I still think my memory is not playing me false when it tells me that Jen was waiting outside and kissed me on the cheek as I left to try my luck on the road to Val d'Isère.

Once out of Switzerland and into France, the hitchhiking improved enormously. I'd be hard pressed to say who drove faster, the French or the Germans, but the French were especially frightening after dark. They seemed to believe that no speed reduction was necessary in towns or villages, provided you blinked your headlights (just once) as you approached any intersection that presented itself. It was a mystery to me why pileups were not occurring everywhere, though I did pass a few. Val d'Isère proved to be everything I had been led to expect from tourist brochures. The switchback roads that wound through the spectacular French Alps were unforgettable. Arriving at the CIM club, I found Guy already there. As a friend of Ian McLaren's brother, he had quickly been made to feel welcome. As Guy set about making introductions, I found myself confused by many of the questions and comments addressed to me by the friendly crowd. My French was still not up to rapid conversation. I became aware that Guy was trying to catch my eye with meaningful looks. As soon as we were alone, he explained that my brother, in order to land another CIM job, this time as a sailing instructor in the Balearic Islands had, with a logic that still escapes me, burnished his own credentials by boasting that his younger brother was a famous sailing dinghy racer who was likely to win a place on the Canadian

Olympic team. For my brother's sake, I would have to maintain the fiction. This would prove a considerable strain.

The CIM accommodation and food were simple and excellent. The crowd of vacationers were fun to be with, and Monsieur Constant was here too. As in Paris, he seemed bent on making our stay as agreeable as possible. It was now the second week of August, and the resort was closed for skiing except for a single Poma lift on a nearby glacier, which closed at noon when the surface snow became too slushy. But at CIM, the partying was non-stop. A magic show was demanded from me every night. I performed in atrocious French, but no one seemed to mind. One night, there was a fancy-dress party made memorable by Monsieur Constant, who did a brilliant turn as a dirty old man with one arm. His actual arm was hidden by loose clothing, and he had positioned his hand to allow the middle finger to emerge out of his open fly. His antics—like approaching a diner (generally a pretty girl), curling *the finger* around a fork and walking away with it—was hysterical to all.

When the time came to say goodbye, there was still, for me, one item on the European itinerary. Guy and I once again split up. We would see each other again only in the departure lounge at Heathrow. My next destination was sixteen hundred kilometres away on the northeast coast of England.

Whitby was the Yorkshire fishing town where my father's roots ran deep. Despite being brought up in Ottawa (I was born in Cambridge, but the family emigrated to Canada nine months later), I was taken to Whitby on many childhood visits, where I always stayed with my grandmother. I have always felt cheated about my birthplace as my brother had been born in Whitby, and somehow this seemed to give him a special status that I lacked. I hadn't known my grandfather well; he had been a severe man I felt somewhat afraid of. He had died in 1956, but I got to know more about him when I had the opportunity to type up the numerous letters he wrote to my father after he moved to Canada. My grandfather was a man of considerable intellect with a particular interest in geology. The letters had been full of enthusiastic geological discussions sent to my father while he was doing his early mapping in the Rocky Mountains for the Geological Survey of Canada. Although Grandad had been too poor as a boy to receive an academic education, he became, in old age, the curator of the Whitby Museum, which celebrated the area's geologically famous Jurassic seaside cliffs and scar, home to rich fossil beds. Still a major tourist attraction, the rapidly eroding cliffs and wide intertidal rock flats (the scar) yield dinosaur remains and abundant molluscs. The latter include beautifully coiled and segmented ammonite shells (analogous to the nautilus still found in tropical waters) as well as the straight cone-shaped shells known as belemnites (now extinct). Local legend has it that St. Hilda, whose ruined seventh-century abbey is the iconic landmark on Whitby's cliff top, cured a plague of snakes by causing them to hurl themselves off the cliff. They then turned to stone. Some coiled themselves as they died becoming the ammonites. Those that remained straight became the belemnites. My father undoubtedly benefited intellectually from my grandfather's geological enthusiasm. Materially, he owed a great deal to the generosity of a wealthy aunt who ensured that, unlike his father, he received a top-notch education. It gives me great pleasure to think that I was lucky enough to continue the line and become the third generation to study and love geology.

Whitby has affected my life in countless ways. Granny had a huge influence on all her grandchildren, and each of us (me, my younger sister, and my older brother) would independently stay with her on our visits to England. We loved her dearly. In appearance, she was quite large and, to my eyes, bore a striking resemblance to a storybook gypsy. She was frequently swathed in cigarette smoke (she favoured Camels) and long, flowing dresses. With her hooked nose, piercing dark eyes, and penchant for oversize earrings, you could easily imagine her telling people's fortunes or holding seances in darkened Victorian parlours. In fact, this image was not far from the truth. She knew how to tell fortunes and often did so. When I revealed my interest in magic, she taught me the arcane meanings of every card in the deck and the moves that had to be made to tell a successful fortune. The spirit world and ghosts were never far away when I was visiting Granny. She apparently believed there was a poltergeist in the house, and whenever any small household item could not be found easily, she blamed it on this mischievous spirit. I was never too sure about whether she was joking. The whole town was famous for its ghosts and their stories. It was probably no accident that Bram Stoker's 1897 novel, *Dracula*, opened with the Count arriving in England on a ship that was wrecked one dark and stormy night on the beach at Whitby. Predictably, Dracula still plays a tacky, but significant, role in Whitby's tourist trade.

And now, I feel obliged to tell you my own ghost story. Are you sitting down? Please make yourself comfortable. First, a word about 17 Bagdale, my grandmother's house. It was a tall, narrow building, just one of the many row houses lining that street. The lowest floor contained the kitchen, the second floor the living room and a separate dining room, and on the third level were Granny's bedroom and the bathroom. I slept in a small bedroom on the fourth floor. My windows looked out over the town, and I could see the ruined abbey perched on the clifftop, a place widely known to have no shortage of ghosts. I loved the continual cries of the herring gulls, which ever since have been synonymous for me with Whitby. They also added a lot to the spookiness of the setting, especially in the mind of a child.

Outside my bedroom door was a small landing where stairs led down to the lower floors. There was also a narrow door with a stiff latch, which opened onto an upward flight of stairs to the first of three attics. Because the door was ill-fitting, the latch was held firmly in place under pressure. The door could not be opened without first leaning your shoulder hard against it to relieve the pressure while simultaneously lifting the latch with your free hand. The same procedure was needed to relatch the door when closing it. The attic itself was a jumble of dusty old boxes and chests full of fascinating junk. I once found an ancient sword in a worn leather scabbard, which I was allowed to take home to Ottawa with me. Another time, it was a six-foot-long pickled marine worm in a large jar

On one visit, when I was twelve, Granny asked me one morning to pack up some of the unwanted junk in the attic in readiness for pickup from the backyard. Working alone, I had to use both arms to carry a full load down the narrow attic stairs. Naturally, I did not stop to close the door onto the landing. This would have required setting down the load and wrestling with the latch. Continuing down the remaining flights of stairs to the second floor where the back door opened onto the yard, I could hear Granny working in the first-floor kitchen and smell the grilled herrings she was making for lunch. Dropping my burden outside the back door, I returned at once to gather another load.

As I came up onto the fourth-floor landing where my bedroom was, I found the door to the attic stairs closed and latched! Startled and beginning to feel more than a little scared up there alone, I dashed down the stairs shouting, "Granny! Granny! Are you in the kitchen? Did you come upstairs?" She shouted back, asking me what was wrong. I held my peace. It was obvious she could not have come up to the fourth floor in the few seconds it had taken to deposit my load. I would have seen her. I decided that explaining myself might upset her. Making up some excuse, I returned to the fourth floor and nervously approached the closed door. I examined it minutely even though there was nothing to see. I tried to unlatch it, but the latch wouldn't budge. I told myself a gust of wind must have slammed the door shut with such force that it latched itself, then dismissed the idea. There was no way a gust of wind could get to the landing, let alone act on the door with that much force—even if there had been a strong wind blowing outside (which there wasn't). Finally, I leaned hard against the door and freed the latch in the usual way.

Sixty-odd years later, I can still feel my beating heart as I ascended the stairs once more. Willing my terror away was not easy. I knew that what had just happened was impossible. Despite my growing interest in magic and the supernatural (I had already read numerous books about ghosts), deep down I knew perfectly well they did not exist. I tried to ignore the mystery while I gathered up another load. This time, I was careful to do exactly what I had done the first time. Down the stairs I went, my arms once more fully occupied with the junk I was carrying. Did I stop on the landing, put down the load, and close the door? No, I did not. I continued on down, dropped the stuff in the yard, and, heart pounding, rushed back up the stairs. The door was closed and latched as before! This was too much for a twelve-year-old skeptic. I was supremely frightened. Nothing at that moment could have induced me to go back up the attic stairs again.

At lunch shortly after, I never said a word to Granny. That night, as darkness fell, I paced my bedroom floor. My door could be locked from the inside, and I wanted more than anything in the world to lock it before getting into bed. But I knew that if I did succumb and locked the door, I would be giving into my fear, and that would make me even more afraid. I left that door unlocked and, somehow, after what seemed like an eternity of listening to every creak and groan of that old house, drifted into a mercifully dreamless sleep.

During the 1980s when I was working out of the geology department at the University of Cambridge, my father arrived for a visit. He had studied geology at Cambridge both before and after the war and, being by then a geologist of some fame, there was no shortage of people in the department for him to meet. At the end of his last day, we went together to the rather posh restaurant at the top of the Student Union building. After the meal, we sat across from each other enjoying large brandies with our coffee and a rare moment of intimacy. For some reason, the story of the latch on the fourth floor of 17 Bagdale came to mind. On an impulse, I told him about it even though I knew his reaction would probably be dismissive and make me feel foolish. He would laugh it off as a childish fantasy not worth talking about.

Instead, he gave a start and stared hard at me. The expression on his face was one I had never seen before. This is what he said. "When I received word that my father was dying, I flew over immediately. I sat with him all that afternoon and was with him when he died. The only thing he wanted to talk about was the fourth floor. Something was upsetting him enormously, but his voice was too weak for me to understand. He died in fear, and somehow the fear was connected to the fourth floor of 17 Bagdale."

To be a child with Granny in Whitby was to be immersed not only in her own eccentricities but those of her equally eccentric friends and acquaintances. Going to the shops with Granny was to meet a kaleidoscope of characters whose idiosyncrasies almost seemed to be treasured. Nearly every day, we would foray together into the streets. First came the fishmonger on Baxtergate where Bob, a man with film-star good looks, always went out of his way to take special care of her (although it was the same with all the shopkeepers). Granny was constantly badgering him for a salmon-trout as a treat for her grandson. It was evidently a rare and special delicacy and hard to get. When he kept failing to find one, she berated him unmercifully. I couldn't help feeling embarrassed for her, but Bob didn't mind. It was part of their game, and he did produce an acceptable salmon-trout eventually (which proved to be delicious). It was impossible to move on without being ushered into the back of the shop where his mother, old Mrs. Eglund, resided in an appropriately damp, fishy atmosphere. She wasn't old; she was, in my young eyes, antediluvian. Granny was wrinkled. Mrs. Eglund was corrugated, and toothless to boot. I had to endure a great fuss being made over me and suffer a torrent of gummy kisses while being gripped in a distressingly long embrace.

Then there was Mr. Brown, the greengrocer, recently released from jail after being charged with selling stolen hams. We lined up in a queue of customers to shake his hand in congratulation. Each person, Granny included, had the same basic message: "So glad you're back out of jail, Mr. Brown! So unfair of the police to charge an honest man like you!" Later, after his shop burned down, he was charged with arson and insurance fraud. He had been seen watching the flames to make sure the building was well lit before calling the fire brigade.

Next was Mr. Bradley, the butcher. Large, rotund, and red-faced, he was unusually taciturn for an acquaintance of my grandmother. He said almost nothing as he served his customers, but Granny would not be put off. She waited impatiently to be served, frequently demanding, "Now then, Mr. Bradley, what have you got for us today? We're in a hurry." This, of course, was an act. Mr. Bradley's eyes would ignore her, and he would say nothing until he had finished serving the others. Then, with ostentatiously slow and deliberate movements, he would wrap up Granny's order, all the while ignoring her repeated demands, "Come on now, Mr. Bradley, what have you got for us?" Finally, and seemingly with great reluctance, he would pass over the brown paper packages of sausages and lamb chops, and looking at Granny straight in the eye, he would lean forward and huskily whisper, "If I was you, Mrs. McLaren, I would try Hot Brandy."

"Finally! Come on, Patrick. Thank you, Mr. Bradley." And she would hustle me out of the shop as if in a rush to get back home, only to drag me through the door of an unmarked, windowless building across the street. There, she marched up to a grinning girl behind a counter and announced, "Put a shilling each way on Hot Brandy!"

Granny loved the horses, and every Saturday afternoon we sat together to watch the races on television and learn if Mr. Bradley's tips were on the money. More often than not, they were. I learned, perhaps a little too well, the joys and pitfalls of gambling. The phone was upstairs in Granny's bedroom, and my job was to run up and down from the TV to call in Granny's last-minute bets to the bookie's office. One afternoon, I got reckless and placed my own bets, which were received with a giggle from the girl, who had come to know me by then. On our visit to the bookie the following Monday, Granny swept in, saying "What do we owe you? How much has my grandson lost?"

"Don't be daft, Mrs. McLaren," came the reply. "We didn't pay a blind bit of notice to the bets your grandson was making."

Whitby was where I started my love affair with the sea and its coasts. The cliffs, the waves pounding on the shores, the cries of the gulls, everything about the place spoke of the sea and its relationship with the land. Whitby was where I first became fascinated with the relentless power of coastal erosion. I could plainly see the tombstones that were doomed to fall next from the clifftop surrounding St Mary's, a twelfth-century church. How long will it take, I wondered, for this medieval landmark itself to be eventually swallowed up by the sea. Little did I know then that I would be focussing my life's career on answering such a question and devising methods to protect shorelines from the relentless eroding power of waves.

Whitby also had an intimate connection with Captain Cook, who had become a hero of mine after reading his diaries. The romance of exploring the world before it was known has remained with me all my life. Cook started out as an apprentice on the colliers that took coal from Newcastle and delivered it all down the east coast. He had lived in Whitby at various times, and the Pannett Museum has a notable collection of Cook memorabilia. The *Endeavour*, his ship for the first of three voyages around the world, was built and launched in Whitby in 1764. His statue on the west cliff overlooks the harbour. A replica of it used to stand on the waterfront of Victoria, BC, opposite the aptly named Empress Hotel quite close to where I now live.[3]

Both my grandparents had deep connections with Whitby's fishing community and its lifeboat station, whose members were revered up and down the coast. The storms in the North Sea are wild and feared, and the lives of the fishermen have always been hard and dangerous. Manning a lifeboat, always done by volunteers, is, if anything, more dangerous. When I first got to know the fishing community in the late 50s and early 60s, fishermen never wore life jackets, nor did they know how to swim. It was accepted that if you ever found yourself in the frigid waters of the North Sea, a life jacket or the ability to swim would simply prolong your suffering. The local fishing boats, known as keelers, were usually less than fifty feet long and had little in the way of guardrails. They did, however, carry a small sail aft of the wheelhouse to help steady the vessel in the waves.

Although the fishing community was comprised almost entirely of men, Dora Walker was a famous exception. A close friend of my

---

[3] Sadly, in protest to Canada's colonial past and racism towards First Nations, the statue was toppled and dumped into the harbour on July 1, 2021 (Canada Day). Cook lost his head in the process. Protest is undoubtedly justified, but Cook's role in such problems seems farfetched. Quite the contrary actually. Cook's assessment and respect for indigenous peoples around the world are clearly evident from his diaries and he both understood and worried about the effects his explorations might have on their future.

grandparents, she was the only woman skipper to hold her licence in the North Sea. She had fished with a gun in her belt throughout World War II and acted as a pilot for boats going through the minefields. She aided in many rescues at sea, and I have two of her seafaring books, *Freemen of the Sea* and *They Labour Mightily*, signed by her with notes of friendship to Granny. She retired in 1953 to become Honorary Keeper of the Whitby Museum among many other honours and died in 1980 at the age of ninety.

One important family connection was with John Robert Storr, who had been born into one of the largest and oldest seafaring families in Whitby. He is pictured, along with three of his sons, in the frontispiece of *They Labour Mightily*. My father had known him in his own childhood. The connection must have been powerful. I remember receiving small Christmas presents from Skipper Storr as a child in Ottawa. This when he was probably as poor as a church mouse after fathering twenty-two children of his own! The story was that they had all become fishermen, were all God-fearing, and never drank, smoked, or swore—except one. His name was Banger. This black sheep of the family was never to be mentioned. He drank, he swore, he never went to church, and he had one wife on the East Cliff and another on the West. To make matters worse, he was the best fisherman of the lot. An inveterate risk taker, he was notorious for bringing his boat, the well-named *Provider*, across the dangerous harbour mouth bar when tides were perilously low or conditions too rough for other boats to follow. With such daring, he always got his fish to market first and so got the day's best price for them.

I had only heard of Banger. My relationship was mainly with another of the brothers, Matty Storr, the owner of the *Pilot Me*. From the time I was eight to about thirteen, whenever I was at Whitby, it was an essential part of the visit to go out on the *Pilot Me*. I was never sure why. I imagine it was more an expression of the respect the Storr family had for the McLarens than a way to give me experience of the seafaring life on the waves of the North Sea. In any case, it was clear that my participation was expected. I was not to let my end down.

Over the years, I got to know Matty well, along with his companion and brother George. George, rather unluckily for a fisherman, suffered from chronic seasickness, but I soon learned that no seaman ever laughs at seasickness. Only landlubbers pretending to be tough do that. Both Matty and George were short, stocky, weather-beaten men, exceedingly friendly but almost impossible to understand. They spoke with a thick Yorkshire accent in an idiom that was essentially a separate dialect. On Sundays, Granny always insisted that I go by myself to the morning service at St. Mary's Church at the top of the 199 steps that led up the East Cliff. "Tell Matty I'm too old for the steps," she would say. There, I would meet Matty, and we would sit, not in a pew but in an isolated, high-walled, separate box that contained padded benches. As soon as the sermon started, Matty would pull out a packet of sweets and whisper, "eez a long-winded boogger, 'ave woon o' these."

At some point during every visit to Whitby, my grandmother would answer the phone and then call me to the receiver to talk with Matty. Doing my best to follow his dialect, I would learn that I was due at the docks at 2:00 a.m. the next morning. There, I would find the *Pilot Me* getting ready to go out. It was exciting to walk by myself in the dark through the narrow, usually wet, cobbled, and deserted streets of Whitby to the dock in the outer harbour, already busy with the hustle and bustle of fishing boats getting ready to go out. Although I enjoyed these excursions on the *Pilot Me*, I was simply a young boy being treated as a guest for a relatively short trip, lasting from early dawn until early evening. I wasn't actually expected to do anything. I got used to the roll of the boat, getting seasick only occasionally, and enjoyed seeing all the various types of fishing that took place: cod hauled in on lines, lobster pots collected and replaced, and the best of all from my point of view, the trawling.

Magic Travels

I had never seen such an abundance of life torn up from the bottom and watched in fascination when the net was released and its contents fell out across the deck. The bounty was rapidly picked over for whatever could be marketed. Squid and octopus, sole and other flatfish, and scallops dominated the catch. Anything not wanted was tossed back over the side. It was no surprise to me when the North Sea started to suffer major catch declines in the 1980s.

On one visit with Granny, there was a dramatic and unexpected change of plan. The usual phone call from Matty summoned me to the dock at 2:00 a.m., and when I made my way there next morning, I found the usual organized chaos of men getting ready, lobster pots and lines being loaded, nets being sorted through, and crates of gear being thrown on board by deckhands, but I couldn't find Matty. It was an eerie scene as figures appeared and disappeared in and out of the floodlights. Faces were hard to recognize, but Matty's was not to be seen. In fact, I couldn't see the *Pilot Me* either despite walking the length of moored boats several times.

A lot of activity was taking place on the *Provider*—Banger's boat! As I walked past for the second or third time, I heard a loud voice emanating from a shape that could have been Matty but wasn't. "As anywoon seen yoong McLaren yet? Ee's meant to be 'ere by now." The shape stepped into a beam of light, and I was facing the infamous Banger. He didn't even begin to look like Matty although he had the same short, squat figure. It was his face that distinguished him. How can I describe that face? It was, quite simply, evil-looking, both bloated and mean, and the eyes were narrow slits that inspected me with no sign of welcome or emotion. "Coom on then, lad. Don't keep us all waitin'. Better get aboard."

I was stunned. What could this mean? I couldn't go out on Banger's boat, let alone with Banger on it. Matty and George, not to mention the whole Storr family—and my own—would be scandalized. I knew that any prospect of living a good and moral life would be shattered forever if I accepted his invitation (or was it an order?). "Thank you," I said, and climbed down the slimy ladder to the deck.

I spent a full three days aboard *Provider*. Although it was never spelled out to me, I knew I had been kidnapped. But why? I could only conclude that this was Banger's way of thumbing his nose at the rest of the family. The shock value of absconding with young McLaren aboard his boat would completely disrupt the respectful ritual established with Matty and the *Pilot Me*. Whatever Banger's motivation, the usual ritual was definitively disrupted. And what a three days they were! It began with standing in the black of night beside Banger in his wheelhouse as we steamed through the dawn on our way to the fishing grounds, but I wasn't to be an onlooker this time. He gave me the wheel and showed me how to steer a compass course. He gave me a lesson on how his depth sounder worked. It would be a museum piece today with its revolving arm that left burn marks on a piece of paper in response to the sound signal bouncing off the bottom. But it worked well enough. When the marks became disturbed, the presence of a school of fish was indicated.

I don't recall seeing a similar device on the *Pilot Me*. The style of command on the *Provider* was also different. Unlike Matty, who worked on deck like any of his crew, Banger never seemed to leave the wheelhouse. He shouted his commands through a side window that could be lowered with a leather strap. I was impressed by the fact that he wore carpet slippers in the wheelhouse. They were his symbol of authority. Banger's job was in the wheelhouse, not on deck. In the covered cockpit of my own sailboat many years later, I always wore slippers and knew that Banger was right. They did give you authority.

The difference from being a mere guest on the *Pilot Me* was profound. I quickly understood that I was expected to work, which I had always been eager to do. The *Provider* was a little bigger than the *Pilot Me,* but with eight or ten on board, the accommodation in the fo'c'sle was cramped. It was a small, dark space with a pervasive smell of fish in every nook and cranny. A small, pot-bellied stove provided just enough heat to make the air wet and muggy from all the clothing that was hung there to dry. A table in the middle was surrounded by two benches that were also used to climb into the berths. Access to each berth was through a small, grimy, oval-shaped hole. It was only possible to get in by launching yourself headfirst into the coffin-like enclosure, after which a series of contortions were required to get your legs inside. Once installed, you were totally safe from the unceasing rolling. There was no possibility of falling out. Perhaps fortunately, it was always too dark to see the exact state of the pillow and blankets that remained in the berth regardless of who happened to be occupying it. The smell of fish on the bedding was powerful.

Three days passed quickly, but I found time to make friends with the cook. I couldn't believe the amount of salt that he added to the water when boiling potatoes, but they were the most delicious I had ever tasted. I spent many long hours in the hold fighting seasickness, while large baskets of cod, still dripping, were lowered down to me to be packed in crushed ice. I learned that whistling was a serious offence. It guaranteed storms. I continued to struggle with the Yorkshire dialect, but by the end of the second day, I began to recognize the swear words being directed at me—though never maliciously. I also got a good laugh from the men when I discovered the hazards of peeing to windward.

I was returned safe and sound to my grandmother. Somehow, word had got through to her that I would be gone for three days, and she hadn't been overly worried, despite Banger's reputation. That voyage unquestionably cemented my lifelong love for the sea.

Now, I was hitchhiking from Val d'Isère to be with Granny and experience Whitby one more time. It would be the last time I saw her. She died in 1971 before I had the chance to visit again.

CHAPTER 7:
OPERATION BOW-ATHABASCA (1966)

From one point of view, the European experience had been a bit of a failure. The travel part had been fine, but had it helped me make up my mind about what to do with my life? Did I now have an overwhelming insight into the career path I should take? Short answer? No. My high-school results had been good enough to get me into Queen's University, and my parents, more precisely my father, had made it clear that university was the only acceptable path for me. I can still hear him saying, "In life, doors open and close. If you don't go through the open doors, they will soon close, and you may never get to open them again." His reasoning had been impossible to dispute. It was settled. My belongings were loaded into the family car, and I was driven to Kingston and deposited in front of McNeill House, the oldest of the university's four men's residences. I wondered why my mother was wiping away tears as the car pulled away. We had never been a huggy-feely sort of family, and it wasn't until much later it dawned on me that her tears had been for me.

Queen's, at that time, had a notably brutal initiation week for first years, even if the rules had recently been toned down a fair bit after a number of deaths. The most notorious ritual was when future engineers competed to climb up a greased pole and snatch a Queen's tam perched on top. The only way to manage this was for hundreds of students, also covered in grease, to climb on top of each other, resulting in a gruesome situation for those at the bottom. In the mid-70s, the university tried to clean up the image of what we called Frosh Week by renaming it Orientation Week, but the rules, such as they were, stayed much the same, as did the purpose, which was the ritual humiliation of the new students by their elders. In my year, the engineering students, for starters, had their heads shaved and their hands dyed bright green. I never saw anyone looking happy about this, nor did I see any point in being pelted with chicken guts or forced to do push-ups on command. Abandoning students, male or female, handcuffed to a streetlamp in their underwear was not my idea of a chuckler. Women seemed to be especially popular targets for debasement.

Frosh Week wasn't the only reason I didn't get along well with Queen's. I had worked moderately hard to get the marks I did in Grade 13 at Glebe, but there was a general belief that university was actually easier than Grade 13. Maybe I took that rumour too much to heart. At any rate, having enrolled in general science, I had to cope with biology, geography, mathematics, physics, geology, English, and philosophy. The last, I think, was a mandatory requirement. All I remember about that class was never having the faintest idea what the professor was talking about. All the first-year courses took place in huge, crowded, impersonal lecture halls. Even when I understood what was being said, I could hardly imagine a duller experience. My roommate at McNeill House was even duller. A lot of beer was drunk, and the level of immaturity was high.

Girls were not allowed into rooms except on Sunday afternoons as long as parents came too. One Sunday afternoon my roommate, normally as staid as they come, apparently forgot that it was Sunday and that his girlfriend and her parents were visiting. They arrived when I was alone in the room, my roomate having gone for a shower. I was chatting politely with them when he got back. Returning naked down the empty hall, he must have heard voices from his room and assumed some friends had stopped by. Flinging open the door, he pranced in holding his penis like a gun and shouting, "Bang! Bang! Bang! You're all dead!" I found it funny, but sadly, neither the girlfriend nor her parents did. The relationship was still new. It did not get any older.

I made it through the year, but only just. My marks were appalling, and I only wanted to get as far away from Queen's as possible. Because geology was the one course I had really enjoyed, I applied for a summer job with the Geological Survey of Canada (GSC) and got lucky. I was selected as a junior field assistant on something called Operation Bow-Athabasca. I would be working on the second of two field seasons aiming to map and interpret the regional geology of some twenty-seven thousand square kilometres of the Rocky Mountain front ranges. It was a landmark project in size and scope and the first time in the geological exploration of the Rockies that a helicopter was used instead of horses. As a result, what would formerly have taken two decades of traditional fieldwork could be accomplished in two summers.

It was already understood that the layers of rock (or strata) seen in the exposed cliffs of the Rocky Mountains, were also buried in the subsurface of the prairies to the east where oil fields were known to occur. The Survey wanted not just to understand when and how the mountains were formed. There was also a huge economic incentive to correlate each stratigraphic rock layer seen in the mountains with the corresponding formations found in drill cores taken in the

Alberta prairies. This required knowledge of the rock type found in each of the layers seen in the mountains together with their age of formation. Age was ascertained using the fossilized remains of living organisms that had died in long-gone oceans, estuaries, and lagoons and been buried in the sediments that later became the rocks that were now so clearly exposed in the mountains. Together, the fossil assemblages and the rock types taken from drill cores could reveal the environmental conditions at the time of their original formation, a key factor in assessing the likelihood for oil to have been formed and stored within the rocks.

My practical geological education started even before my arrival in Calgary, the rendezvous point for the Bow-Athabasca team. My train had no sooner left Union Station in Ottawa when I had the good fortune to meet Dr. Ken North, who was headed west on unrelated business. A friend of my father and head of the geology department at Carleton University, Ken was a Yorkshireman who, it can justly be said, was larger than life. He was loud, opinionated, and almost impossible to stop when he was pursuing an argument (which seemed to be all the time), but my father claimed that Ken was single-handedly responsible for turning the university's geology department from a joke to one of international renown. My father also said that in any discussion (meaning argument) Ken North had one crucial advantage. He had a photographic memory.

Ken always referred to his wife, who must have been long-suffering or exceptionally forgiving, as Clueless. This was not a description. It was a substitute for her name. He was known to enjoy beer-drinking competitions. When challenged to one, always by someone who didn't know better, he would first flick the caps off two bottles and down them together. Then, banging the empties on the table, he would announce, "Right, my two to your one." As this probably suggests, he was always the one who remained standing at the end of the contest. He had worked equally hard to make a mediocre geology department into something to be proud of, but university politics are not as straightforward as drinking contests. Once, he was overheard complaining about how impossible it was to remove an incompetent, but tenured, professor in order to make room for someone better. "The only way that bastard can be fired is to catch him *in flagrante delicto* with the Chancellor's wife. Not once but repeatedly!"

Ken made the two-day journey across Canada into a memorable, educational, and eminently enjoyable interlude. The train had a rather grotty bar car. It was usually filled with a fairly rough clientele, typically miners, lumberjacks, and assorted labourers seeking their fortunes in the expanding west. The air was always dense with smoke and noise, and my ID was never checked. Pipe-smoking Ken insisted on taking me along. He not only seemed to belong there, he filled the space with his presence. His booming baritone demanded attention and his manner commanded respect. I had never seen anything like it. Charming and witty in his entertaining Yorkshire accent, he frequently bought drinks all round and soon knew everyone. He also seemed to know exactly how to make them do exactly what he wanted, which was to listen to him talk about the geology of the landscape as it unfolded outside the window.

And what a story he told! Although today the concepts of continental drift and plate tectonics are widely, even easily, understood (a child can see on a world map how the continents of Africa and South America would fit together like pieces of a jigsaw puzzle), this was not the case in 1966. The founding father of continental drift, Alfred Wegener, died a broken man following the criticism he received after he published the observational evidence (mainly rock types and fossils) for continental drift in 1912. Because a plausible driving force to explain how continents could split apart and move elsewhere could not yet be explained, Wegener's evidence was not only rejected but often ridiculed. It was even called German pseudoscience by some, and Wegener himself was accused of allowing an unsubstantiated theory to run wild. After his death, continental drift became the greatest geological controversy of the twentieth century, and in 1966, the world of geologists was still sharply divided between those who believed in continental drift (which explained so much and, in the end, completely revolutionized geological thinking) and those who didn't. Ken most definitely did.

Today, we are used to videos explaining all kinds of planetary evolution. What Ken was able to do in that bar car as we crossed Canada in 1966 was the verbal equivalent of a brilliant video, explaining the passing rocks and landscape in clear, precise terms that anyone could understand. In the Ottawa valley, he took us through the Precambrian era over a billion years ago when marine algae ruled. This early primitive form of life eventually produced the oxygenated atmosphere that enabled the more advanced forms familiar to us today to start evolving about 600 million years ago. The algae enabled sediment to be bound in layers to form mound-like domes called stromatolites. Thought to be long extinct, the discovery of living stromatolites in Shark Bay, Australia, in 1956, had been a source of intense geological excitement. Even as I write this in 2021, the latest Mars rover, *Perseverance*, is attempting to discover whether similar life forms once existed there (a first step in understanding if they are likely to be found anywhere else in the universe) by coring into three-billion-year-old rocks in a geological setting suggestive of stromatolite formation.

With Ken, wreathed in his pipe smoke and swigging his beer, a single outcrop of rock passing the window was put into the context of colliding plates, continents being ripped apart, the birth of great mountain ranges that had long since eroded away to nothing, massive floods, earthquakes, and moving ice sheets more than three kilometres thick. His audience was transfixed. Ken left no question unanswered, and there were many. It was a virtually non-stop, two-day performance, and I have no doubt there were others besides me in that car who eventually became geologists as a result.

∞

We arrived in Calgary shortly after noon on the third day. The first task was to buy boots for the rigorous field program that was in store for me. Boots are rightly considered the single most important item of geological field gear. Their purchase was the responsibility of each person taking part in the operation. At the shop recommended to me, I discovered that the salesman seemed to know every geologist in Calgary. I followed his obviously expert advice and bought a pair of high-cut boots suitable for snow, shallow rivers, swamps, mud, and, of course, rock and scree. I had been told it was rare for any pair of boots to last more than one summer under the unimaginably tough field conditions we would be contending with. Then I checked into the York Hotel and sought out Dr. Ray Price, the party leader and brains behind Operation Bow-Athabasca. Ray was a heavy-set man, a pipe smoker like Ken, though unlike Ken, he thought carefully before talking, and when he did so, the words came out slowly.

I was lucky enough to act as Ray's field assistant on several occasions. He was unflappable. I never once saw him visibly annoyed or hot under the collar in the face of events that would have sent most people, including me, into a rage. One day, a good deal later in the field season, when the helicopter was fully occupied ferrying other geologists to their various mountain locations, I became his chauffeur. As I drove one of the GSC panel trucks on remote logging roads, Ray, his air photos spread out across his lap, gave me directions. We stopped at outcrops here and there to collect specimens and search for fossils, with Ray making detailed notes at each stop. To my embarrassment, I had bad luck with my driving. I managed to get the truck stuck, as Ken North might have said, *not once, but repeatedly.*

Twice we were able to dig ourselves out, but not the third time. Trying to make a multi-point turn in the confines of a narrow logging road, I backed just a little bit too far over an embankment. I can still hear the CRUNCH! as the vehicle came to rest with its full weight on the chassis. Scrambling out to survey our predicament, we found the truck suspended with all four wheels off the ground. Digging out was impossible. Was Ray upset? Did he curse my stupidity? Not a bit. "Patrick," he said calmly, "I'm sure we can find a farmer with a truck to get us back in business." So off we trudged to the highway several miles away and immediately spotted a farmhouse. The farmer answered the door and did not seem surprised to find two dusty strangers there. After we explained our plight, he told us it was Farmer Joe we wanted. "He's the third farm on the left. He's got a truck with all the gear to get ya out. He'll help ya. But Jesus Christ, watch out for his dogs. Ya don't want to mess with any of them."

A little discouraged by this, we decided to try the next farm on the left first. There we got the same reply. "You need Farmer Joe, next farm down. He'll help yous out fer sure. But you gotta be careful. His dogs are vicious."

At the start of the lane that ran down to Farmer Joe's, we stood together staring thoughtfully down its length. "Well, Patrick," said Ray finally, with only a slight glint in his eye, "I'll just wait here while you see if you can find any help."

I couldn't blame him. After all, it was my fault we were in this fix, and I was quite obviously the expendable one. I set off down the lane. I was still carrying my rock hammer loosely in my right hand and could feel it brushing my calf as I walked. Approaching the house, all my senses on high alert, I became aware of that special kind of silence that suggests all inhabitants, human or otherwise, were out. There was no hint of life anywhere.

How wrong I was. Just as I had begun to relax, a massive German Shepherd sped out from the open door of the nearby barn. Unlike the previous summer's close encounter with two East German canine border guards, there was no intervening fence to hold this one back. His teeth were bared in a wicked display of savage fury, but not a sound came from him. His speed was phenomenal, his intent crystal clear. He was going to tear me apart, starting with my throat. As he reared up for the kill, and without any conscious thought on my part, I swung the rock hammer and smashed it into the side of his head. The blow knocked him sideways, and he crashed to the ground, yelping loudly.

He lay still for a moment, then still howling, fled into the depths of the barn as fast as he had come. I wasn't sure what to do. Had I smashed his skull in? Was he dying? Should I go and check? With horror, I saw that the farmer had now appeared on the farmhouse porch. He had obviously seen the whole episode. I braced myself. It wasn't hard to imagine that he might be far more dangerous than his dog.

"Well, you sure showed him! Haw haw haw!" he laughed. "He won't be trying that again in a hurry! Haw, haw haw! That was amazing! What are doing you here anyway?"

Still shaking, I told him. A moment later, we were in his truck with some lengths of chain in the back on our way to pick up Ray and get our own truck back.

We had established a base camp on the flood plain of the south bank of the Blaeberry River, a westward flowing tributary of the nearby Columbia about ten kilometres north of the town of Golden, BC. For the first few days, the fifteen of us, geologists and assistants, plus Fred the Cook, busied ourselves with getting the camp up and running. On innumerable trips into Golden to pick up assorted odds and ends from the town's one hardware shop, I quickly learned that letting any of the geologists drive was to be avoided as much as possible. They drove with their heads out the window, studying the stratigraphy that was spectacularly displayed in the exposed mountainsides on either side, barely avoiding the logging trucks that seemed to be everywhere. It was amazing they all survived, but after my first two or three rides with a geologist in the driver's seat, I always came up with some reason to take the wheel myself.

Regardless of how busy they were with the preparation work, the geologists always had time for scientific discussions which often evolved into heated arguments. "That's the Upper Devonian Palliser Formation. You can see the contact with the Exshaw." "Don't be ridiculous! It looks decidedly dolomitic to me. It has to be either the Cairn or South Esk. And see that fault over there? That's what's making the boundary confusing." And on it would go, sometimes for hours, while we came to grips with all the intricacies associated with pumps, plumbing, propane water heaters, generators, and electrical circuitry that was necessary for the camp to operate smoothly during the field season. The completed camp, comprised of several trailers and storage tents placed neatly around a central common area, was remarkably comfortable. Besides sleeping quarters, we had hot showers, toilets, office space, and a well-equipped kitchen with attached dining area. There was plenty of room nearby for a stash of aviation fuel in forty-five-gallon drums and a helicopter landing area.

In the diary I kept, I recently found two pages of sketches showing all the formations we were likely to be mapping, some thirty in all. Each was designated with a symbol that identified the rock type and its age. The oldest was late Cambrian (about 495 million years old), the youngest (lower Cretaceous) at only 140 million years. It was absolutely expected that all the assistants would have this information memorized. What I didn't know then, and only discovered while writing these words more than a half-century later, was that it was my father who had named many of these formations in 1955.

When the helicopter arrived, Jim Davies, a greatly admired mountain pilot who was already famous for a number of daring mountain rescues, was at the controls. It was a small, versatile aircraft, a Bell 47 G-3B1, that carried only two passengers and the pilot, all sitting in a row inside a clear plastic bubble. First made famous during the Korean War, the Bell 47 was best known from the popular TV series M*A*S*H. Helicopters were still rarely seen by the general public and flying in one was even rarer. I found it thrilling and remembered how I had dreamed of flying to the mountain heights while on my first train journey into Jasper. Now I was doing this every day.

Jim certainly prevented the experience from ever getting boring. One day while flying was still a novelty for me, I happened to be the sole passenger and was waiting in the seat beside him for the helicopter to warm up before we took off from base camp. On the console in front of me, I noticed a decal with several warning messages printed in red. The last of these was written in highly ungrammatical English, but the meaning was clear—DO NOT FLY BACKWARDS. I leaned over to Jim. "What does that mean?" I shouted over the engine noise, putting my finger beside that perplexing line on the decal.

"Dunno," Jim shouted back. "Never read it before." He then proceeded to lift off and fly perfectly backwards down the length of the clearing before pausing to hover a few feet above the ground. Then he ascended vertically at speed. It took a few minutes to remember to unlock my white-knuckled fingers from the straps of my seat belt.

Jim also delighted in swooping low over any bus on the highway. Fun for us once we got used to it, but I doubt if the startled driver thought so. Each day, weather permitting (about which Jim's word was law), I was flown, usually with one of the geologists, to a selected site, typically on the flank of a mountain. After setting us down, Jim would return to base to ferry another pair to another location. As there was no radio or phone communication in those days, the time and place for pickup at the end of the day needed careful discussion.

Another Jim, Jim Aitken, was a geologist I liked a lot. Tough and wiry, he was a genuine outdoorsman and a mountain climber who prided himself on reaching geological formations that others feared to attempt. It was Jim Aitken who undertook to give us basic survival lessons. He taught me a great deal. We were often roped together in our climbs up and down cliffs as we collected fossils. The days were long and strenuous. Our packs grew heavier and heavier as the day wore on. On one traverse down a steep glacial tongue, I was in front when we came to a deep crevasse. It didn't look all that wide until I came up close to it. Then, it did. I was sure it was too wide to jump, but Jim was impatient.

"C'mon Patrick, you can jump that. We don't have all day."

I inched as close to the edge as I dared. "Okay," I shouted back, "but give me some more rope." He complied. "No, I need more than that," maybe yelling a little louder than necessary. I was sure that Jim was beginning to take me for a wimp. Even if I jumped well, without enough slack rope to make the distance, my fate would be sealed. Wimp or not, I took my time until I was fully satisfied with the amount of slack Jim allowed me. Then I swallowed hard and jumped like hell. I overshot by a surprising margin and felt pleased with myself. Now it was Jim's turn, but our roles were reversed. I had control of the rope. Jim approached the edge, then he stopped.

"Oh," he said, "yes, I see what you mean. It is kind of wide." He paused for a moment. "Better take up all the slack," he said. "No, tighter than that, and keep it tight as I jump." He gave away his anxiety by clearing the distance by even more of a margin than I had. If the crevasse had been only a foot deep instead of what it was, a deep gap in brilliant blue ice that was like gazing into infinity, neither of us would have thought twice about jumping across it.

Jim and I shared a common sense of humour. We both were fans of the British comedy group, Beyond the Fringe, the forerunner to Monty Python that included Peter Cook, Alan Bennett, Dudley Moore, and Jonathon Miller, all of whom went on to have major careers as writers and performers. We could both recite verbatim Peter Cook's explanation of why he had become a coal miner instead of a judge ('e didn't 'ave the Latin) and would keep ourselves in stitches while we dangled from our ropes conversing in ludicrous British accents. It was more this humour than anything else that kept us going one dreadful night when we had to huddle until dawn under a sheltering ridge while freezing rain and snow lashed down. Unaccountably, Jim-the-pilot had not arrived at the designated pick-up, yet we knew him well enough to be confident he would come for us as soon as it was light. Sure enough, just as dawn finally broke, we heard the beautiful chopping sound that would become so familiar throughout my career—a helicopter on its way.

Two climbers had gone missing the day before on Mount Assiniboine, so finding them had been more urgent than picking us up. They had still not been located, and Jim was in a hurry to get back to the search. His wife, Siri, owned and ran Assiniboine Lodge at the foot of the mountain, which catered mainly to climbers and hikers. Jim dropped us off there since it was nearer than the camp and then rejoined the search. Exhausted and hungry, we entered the main hall where breakfast was being served. It was full of hikers. Having just arrived by helicopter and looking as cold, wet, and all in as we must have done, we created quite a stir. Maybe we were more than a little punch-drunk after our miserable night in a blizzard at the top of a mountain, but we couldn't seem to break out of Beyond the Fringe mode. In a loud voice, Jim said, "'ere Patrick, 'ave you ever seen so mooch fookin' bacon?"

In our condition, this seemed tremendously witty. We could hardly stop laughing. The rest of the room felt differently. We found ourselves alone at the buffet while at the other end of the hall a crowd of guests huddled together, staring dubiously in our direction. Fortunately, the missing mountaineers were rescued, as were the guests when, after an outstanding breakfast, Jim arrived back to take us away.

Jim-the-pilot, along with his superb flying skills, had some quirks that required a degree of fortitude on the part of his passengers. One long traverse with Jim-the-geologist saw us deposited halfway up the side of Mount Stephen, one of the larger peaks (3,200 metres) immediately behind the small town of Field, BC. Its looming cliffs dominate the town, and it has been much painted by artists, although as Jim would scornfully point out, "they seem consistently unable to see the stratigraphy." Jim often made this observation about mountain art that was displayed in shop windows. He was always affronted. The lack of geological insight worried him. "How do they not see the stratigraphy?" he would mutter to himself.

We spent a long day, roped together, climbing and collecting until we reached the summit in the late afternoon. By then, our packs were heavy. We were tired, but Jim was concerned about how the helicopter would land in the deep snow. He decided we should stamp down the snow to give Jim-the-pilot a firmer landing pad. This was harder work than anticipated. It took us a full hour of trudging and stomping until he was satisfied with the result. We heard the helicopter coming for us just as we finished. It was a long slow climb, and we watched as Jim-the-pilot achieved the necessary elevation by making repeated circles as he spiralled the machine upwards. Finally he got above us and came down to land.

Did he gratefully settle down on the clearly visible landing pad we had painstakingly prepared? Of course not! This was our punishment for taking responsibility for things that did not concern us. He was the pilot. He decided where to land. And as the cowboys would say, he liked being ornery. He was no doubt chuckling at our obvious consternation as he selected his site in the deep snow right beside our pad. We tied our packs to the outside racks and climbed in while Jim kept the rotors whirling to maintain a little bit of lift. As soon as the seat belts clicked, Jim, without lifting off, proceeded to ski the helicopter off the cliff edge and into empty space. If he was out to give us the ride of our lives, it worked. If he was out to scare the living bejesus out of us, that also worked. The town of Field appeared to be shooting up toward us, spiralling as it did so. The speed of the drop left our stomachs back at the mountain top. Never shy about touting his own bravery, Jim-the-geologist, when relating this story, had the grace to confess, "I damn near shat myself!"

∞

In the mid-1950s, our whole family had shared a field season with my father. This was probably the expedition when he came up with many of the rock formation names I had recently learned. I think I must have been seven. At any rate I was old enough to remember an episode with a cinnamon bear (a subspecies of the black bear) that had begun hanging around the cook tent. One night it visited havoc on both the tent and its contents. It was promptly shot by one of the cowboy-packers. Its massive corpse left a deep impression on me. Since that moment, I have never accepted that shooting any wild creature is justified by the need to ensure our human existence on the planet. Nevertheless, I eagerly took part in the practice-shooting at tin cans that everyone in the camp was required to do. But I have been lucky in not having my pacifist principles put to the test. Shots fired over their heads have always proved sufficient to make approaching bears (including polar bears) retreat (out of danger from me).

Looking back at Operation Bow-Athabasca, I find myself wondering how it was that we got such minimal training in dealing with potentially dangerous wildlife. We spent most of our time in Banff and Yoho National Parks, where carrying guns is not allowed. Today, park signs do at least provide some guidance on how to behave when encountering a bear or cougar. My adventures with Tom McCready in previous summers had left me with a rather cavalier attitude and a conviction that simply shouting and throwing rocks would always be enough. Tom was highly skilled in this technique. Once, when a bear was a bit too nonchalant about removing himself, he just laughed and exclaimed, "Jeeesus! They don't even scare anymore these days!" Then he hurled another rock.

Tom also assured me that bears would always leave you alone when you were on horseback, and I have found that to be the case. We weren't on horseback with Operation Bow-Athabasca, yet bear training, if there even was such a thing at the time, seemed to be of no importance to anyone. As far as cougars were concerned, I had already learned what their scat looked like, but I have only once seen a live one. That was in 2018 just outside my house on Saanich Peninsula near Victoria, BC.

Bears were common. One memorable encounter occurred when I was with Don Cook, a PhD candidate who was completing his research with us and would go on to a distinguished career as a GSC geologist. Don was from northern Ontario and had had enough experience with black bears to fear them. I had been his personal assistant for several weeks, and we had moved from the base camp to a campground in Yoho near the foot of the breathtaking Takakkaw Falls. I had my own tent, while Don and his wife lived in a trailer with their new baby. It was always exciting when Jim put down the helicopter among the tents. The noise and the wind generated by the rotors quickly brought the campers scrambling out to find the cause of such a commotion. Don and I would stroll nonchalantly out as though our limousine had just arrived.

We had just disembarked from the helicopter in a small clearing in a heavily wooded area and were watching Jim depart, anxious as usual about the proximity of trees that seemed just a little too close for comfort to its whirling blades. That concern quickly faded when, turning to start our traverse, we narrowly missed putting our feet into a pile of fresh, steaming bear turds. We couldn't determine the direction the bear might have taken and began to work our way through the bush into a stream valley where we knew we would find the outcrops we were looking for. An instant later, we found ourselves only a few yards away from two grizzlies. They were busy swiping huge chunks of earth out of an ancient riverbank and hadn't noticed us yet. I presume they were hunting for marmots as the bank contained numerous burrows.

As we stood frozen to the spot, some ursine sixth sense must have kicked in because they stopped what they were doing and with one accord turned their great heads and looked directly at us. "C'mon Patrick, run for it!" hissed Don and took off. I turned tail and followed. What a mistake! Running through bush is just about impossible. It is out of the question to look anywhere except where you are going. Taking even a second to lift your eyes to any other view, let alone try to look backwards to see if you really are being chased, is not an option. The probability of a serious fall is too great and, in those circumstances, could lead to being eaten alive.

Panic was inevitable, and as everyone knows, carnivorous animals get excited when their prey is panicking. So the situation seemed as dangerous as it could possibly be. To make matters worse, we were in a Catch-22. To keep running made falling, sooner or later, almost certain, but to stop was absurd. The bears could have been right behind us. In the end, however, exhaustion takes over, and you have to stop whether you want to or not. Which is exactly what we finally did. Panting heavily, we looked sheepishly at each other. There were no bears in sight. We had been lucky. We were equally lucky not to have injured *ourselves*. That could have put a successful outcome of the whole project in jeopardy.[4]

∞

Don and I worked well together. Since we were away from the base camp for an extended period, we were often able to do without the helicopter and spend more time on specific details associated with Don's PhD research. This included using road cuts to measure the orientation of what are known as slickensides. These are striations formed on rock surfaces when crustal movements slide one mass of rock against another, resulting in faults. The directions of the striations, when plotted in three-dimensional space, reveal how the rock masses must have moved during the processes of mountain building.

On the last day of the field season, Don planned a particularly ambitious traverse. With the help of Jan, Don's wife, we drove two vehicles to a locked gate at the entrance to a logging road, having picked up the key at the park office. Planning to end our traverse at the gate, we left a vehicle there. Jan then dropped us off where we wanted to start the traverse some miles away.

It was a long, long day during which we covered about twenty miles without trails. We climbed up the sides of mountains, crossed valleys, and forded streams, splitting up whenever we needed samples from both sides of a valley. By late afternoon, our packs were once again overloaded with rocks. We were wading a stream when I glanced up its course and spied an animal I couldn't readily identify. It too was crossing the stream but with a curious, loping gait. It was bigger than a badger but smaller than a bear. I felt certain I was looking at my first wolverine or, as it was commonly called, a Carcajou. As a child, I had been fascinated and impressed by

---

[4] In 2021, I invited Don Cook to read this chapter. He said he enjoyed it, but, for the record, he would like me to put in a footnote that he has no recollection of this event ever occurring.

Rutherford George Montgomery's classic wilderness adventure story, *Carcajou the Wolverine*. Its exciting cover showed a ferocious, yellow-eyed creature circling dangerously around a much larger but plainly terrified wolf. The opening paragraph said it all: "He had nothing to fear, for he was king, and every killer from the cougar to the grizzly knew and feared him. His reputation made him king. Hated and feared, he freebooted and slaughtered alone. Glutton, skunk bear, wolverine or Carcajou the killer, each name fitted him."[5] No wonder I had loved that book!

These memories came vividly to mind as I watched Carcajou himself meander toward us with that distinctive shambling way of walking, apparently still unaware of our presence. Though over the years I had gained a more realistic knowledge of wolverines, I had never seen one before. I knew they were the largest of the Mustelid family, which includes otters, mink, and badgers, and they had a reputation for boldness and savagery that Montgomery had not exaggerated. I had read a documented account of a wolverine taking on a polar bear. And winning. I knew too that they are loners and extremely shy, which is why they are seldom seen. Remembering my encounter with Farmer Joe's dog, I reached for my rock hammer and warned Don, "We should be careful, these things can be dangerous." I took his rock hammer from the back of his pack and handed it to him. As Carcajou the wolverine came nearer and nearer, still seemingly oblivious to us, I was struck by how beautiful the creature was with its dark snout and a white V of fur surrounded by a luxuriant brown and black coat. Its eyes were sharp and intelligent. Its claws, however, were also sharp and formidable looking weapons.

Soon Carcajou was directly in front of us. There was no doubt that he (or perhaps she) saw us now but seemed not in the least surprised, merely staring at us with evident interest. There was not

---

5   Rutherford G. Montgomery, Carcajou the Wolverine, (Caldwell: Caxton Press, 1936), ISBN 10:0870044036 ISBN-13:978-08700440381.

the slightest indication of fear or antagonism. Just the opposite. This animal was clearly in a friendly mood. We stood with our heavy packs on, hammers at the ready as the realization sank in that there was absolutely nothing to fear. It was an extraordinary feeling, and we both felt it. We were communicating with this utterly wild animal that had taken the trouble to come over and say hello. Entranced, we removed our packs to better enjoy the encounter. Carcajou spent some time examining them carefully, then sniffed around our legs. I am sure we could have petted him. We didn't.

It was getting late by now, and we needed to move on, so in the end it was us that left Carcajou. Looking back as we disappeared into the trees, I had a last glimpse of the only wolverine I have ever seen watching us depart. It would be too easy perhaps to say that Carcajou looked sad, but I was convinced that this was the case.

Our traverse came to an end when we reached the logging road. The truck was another hour's walk away. Flipping a coin determined that I would be the one to get it. I didn't mind. Don had work to do with his notes, and with the packs in his care, it wasn't unpleasant to continue walking unencumbered. As I set off, it occurred to me that at this time in the late afternoon all sorts of animals might well be present. I began to sing and whistle to give warning that I was coming. The result was the opposite of my intention. Without warning, a large bull moose stepped silently out of the bush onto the road just yards in front of me and turned its head to inspect me. Apparently satisfied, he began to amble down the road in the same direction I was heading. For the next half-hour we walked together in single file. Every so often he would look back over his shoulder as if reassuring himself that I was still there.

He was an impressive animal with a massive rack of antlers. I decided to take the precaution of arming myself with rocks just in case. With a smooth, comfortably sized rock in each hand, I decided that if he did charge me—though I didn't really believe he would—I would hurl the rocks as hard as I could and then run like hell. The only other options were to go back the way I'd come or sit down and wait for him to get well ahead of me. With darkness drawing on and Don waiting for me, neither was feasible.

I had nothing to fear. As stealthily as he had appeared, the moose abruptly turned and disappeared back into the woods. The road was taking me around a steep rocky point when I heard sliding rock and snapping brush to my left. I walked to the edge where the valley side sloped away steeply. Peering down, there was the third animal of the day. Like Carcajou and the moose, this one was also close. Our eyes met and locked on each other. It was a grizzly. As I stood there, he began to lever himself off the log he was standing on and brace for a lunge up onto the road. That moment was frozen in time for the rest of my life. What I felt was not exactly fear. With an odd kind of detachment, I reflected that this enormous bear, in a few seconds, was going to be right beside me. What would that mean? Would he be willing to share the road with me like the moose? That was hard to imagine. And what if he was hungry? Or simply territorial? My mere presence might seem like an insult. The consequences for me were likely to be the same either way. At that point, conscious thought stopped. In full cowboy mode, I shouted at the top of my lungs, "HEEEEEEEAAAAAGH!" and as hard as I could, I flung one of the rocks that was to have protected me from the moose and which, for some lucky reason, I had not yet discarded.

Now I should explain that my throwing skills are extremely limited. I put this down to being mixed up about whether I am right-handed or left-handed. I write with my left hand and do plenty of other left-handed things. But I cannot throw with my left hand. Unfortunately, my right hand isn't much good either. Every year in high school, we were put through the hexathlon, a series of eight graded athletic challenges. I usually got 100 percent in seven of them. In the eighth, competing to throw a ball as far as possible, I always got zero. I couldn't even make the minimum distance needed to score a single. The humiliation was intense. In those sexist days, no one hesitated to inform me that I *threw like a girl*.

Maybe it takes an approaching grizzly to focus the attention properly. If I were a religious man, I wouldn't hesitate to claim my throw was guided by God. The stone flew from my hand and bounced off the bear's forehead, right between the eyes. Surprise, shock, and panic (in that order) registered on his face. Then his back legs slipped from the log, and without further ado, he took off at full speed down the steep slope. I could hear the sound of his passage as rocks tumbled and branches snapped in his headlong panic to . . . well, to get away from *me*. Transferring the remaining rock to my newly empowered throwing hand, I stood at the road's edge listening to the fading sounds of a grizzly bear in flight. I felt like Tarzan and had a strong urge to beat my chest.

As if enough hadn't happened already, there was still more to come. It was already dark by the time I reached the truck. Pausing for a quick pee before driving back to get Don, I suppose my nerves must have been a little on edge. Without warning, the air all around me filled with a loud whirring noise accompanied by a frantic, rhythmic beating. Involuntarily, I jumped straight up and, as I came down, caught sight of a family of ptarmigans taking off in fright at the sudden warm shower. I drove back to pick up Don. "What a walk I had! You're not going to believe this." He didn't make any comment after I told him, but the look he gave me seemed to say, "Crazy bugger! That could only have happened to you!" Then he roared with laughter.

CHAPTER 8:
BAFFIN ISLAND, CANADIAN ARCTIC (1967)

The diary I kept during Operation Bow-Athabasca is a disappointment. Instead of focusing on what actually happened, it consisted mostly of navel-gazing. Having never looked at it again until I dug it out to help me write about that summer fifty-four years ago, I was dismayed to find only the musings of a self-absorbed teenager seemingly incapable of anything but agonizing over what he should do with his life. I felt guilty for wanting to take off and travel the world. I worried about being lonely without my friends, especially Mary, my high-school sweetheart. I dreaded going back to Queen's, but thought I should and fretted about what my father would say if I didn't. I fastened on an invitation to join one of the geologists on a drive to Texas at the end of the field season, and I outlined a half-baked plan to continue from there to South America. I spent many evenings with a teach-yourself Spanish book I picked up heaven knows where. The diary was embarrassing to read. I come across as a spoilt whiner with no idea of what a privileged life I was actually living. And, in the end, the tortured musings and crazy plans came to nothing. I succumbed to what I wanted to do least and went back to second year at Queen's.

This time, I at least rejected the idea of staying in a residence. Instead I found a room of my own in a boarding house run by a Mr. and Mrs. Wiskin. It was in a rundown part of town, and the Wiskins were a retired couple who made ends meet by renting three small spare bedrooms to impecunious students. The bedrooms, the Wiskins' as well, were all on the second floor with one bathroom for the lot of us. Let's call the arrangement cozy. Mrs. Wiskin ran a tight men-only ship. Breakfast sharp at seven, supper at six, and no female visitors at any time. On Thursdays Mr. Wiskin dressed up in his Odd Fellows regalia and dined with us before his missus let him out to his Meeting Hall. He was a kindly man and extremely proud of his costume, which featured a conical (and comical) hat, embroidered vest, a sword (real as far as I could tell), and a host of arcane medals and ribboned medallions. He was unable to explain exactly who the Odd Fellows were or what they did. Perhaps he had sworn an oath of secrecy, but I am almost sure he really didn't know.

One of my fellow lodgers was Keith Dorland, who was taking English and wanted to be a writer. Quiet and intense, he took life seriously. He eventually settled in London and had several plays performed. Coincidentally, I ran into him again many years later in the London tube. The other, John Kreeft, was Keith's polar opposite. An extremely bright and extroverted med student with no shortage of ego, John could also be an annoying know-it-all, but he had a macabre sense of humour. He once took great pleasure in introducing me to the cadaver he was dissecting, manipulating its arm to offer to shake hands.

I liked both my housemates, but I knew that secretly Keith disliked John and suffered him in silence. What we all genuinely shared was a love of music. I had studied piano throughout my childhood and high-school years and had become good enough to play (after untold hours of practice) the odd Chopin Polonaise or Schubert Impromptu. I had also played flute in the Glebe band and taken part in any number of musical get-togethers with friends. But I knew I was only a mediocre performer. Playing an instrument had been drummed into me as something important, and I had practised diligently enough mainly to please my mother. And, after all, she was right. I never got good but playing gave me a lifelong appreciation of classical music, and throughout this second year at Queen's, I bought classical LPs whenever I had a little extra money. Almost every afternoon after classes, I would lie on my bed and listen to whatever composer I was into at the moment. I never moved on to a new composer until I felt that I knew at least some of the music of my current favourite intimately. I started with Bach's Brandenburg Concertos and the Preludes and Fugues. Next came the Beethoven symphonies, then Handel's Messiah. Brahms, Chopin, Schubert, Mahler, Sibelius, and Saint-Saëns followed. I made it as far as Prokofiev, Shostakovich, and Stravinsky before deciding that I would never understand or enjoy classical music past the 1920s.

John Kreeft had studied the French horn and was keen on demonstrating how it was put to such good use in the last movement of Sibelius's fifth symphony. Keith Dorland had never been introduced to serious music but became an enthusiastic convert. I remember how excited he was when he first heard Saint-Saëns's fourth piano concerto. Our listening sessions and discussions became an integral part of our friendship throughout that year. They were certainly more educational than most of my lectures.

The university year passed in a state of great and unhappy confusion. I spent my time hanging about with friends, skiing at every opportunity, avoiding lectures when I could, and daydreaming of travel. My closest high-school friend, Don Hindle, was over two hours away in Ottawa, taking English at Carleton, but somehow we still saw a lot of each other. University was proving to be intolerable for both of us. Together we decided to take off and travel as soon as the academic year ended. The only question was how. Then one of us, I don't remember who, had a brilliant idea. Why not get a sailboat and see the world that way? There were just two problems. Apart from my experiences in dinghies, mainly on placid Dow's Lake in the middle of Ottawa, we could not have been more ignorant of what ocean sailing actually entailed. There was also the little matter of money—we didn't have any.

Such trifles did not dampen our enthusiasm. I read and dreamt about sailing constantly. Joshua Slocum[6], Bill Tilman[7], and Miles and Beryl Smeeton[8] were my heroes, but I pored over the stories of any number of other famous sailors. At one of our get-togethers, Don and I formulated a plan: (i) finish the university year, (ii) use the summer to make some money, and (iii) find a boat in the fall that would take us on as crew. It was clear, straightforward, and childishly simplistic.

---

[6] Captain Joshua Slocum (1844-1909) was the first to sail around the world single-handed. Departing from Halifax in 1895, he took three years on board his thirty-seven-foot, gaff-rigged sloop. His whole life was an adventure, and his book *Sailing Alone Around the World* became a classic.

[7] Bill Tilman (1898-1977) was a military hero, a world-renowned mountain climber, and an extraordinary sailor. His sailing was often just a way to get to the remote mountains he wanted to climb. He wrote many books, one of my favourites being *Ice With Everything*.

[8] Miles and Beryl Smeeton are perhaps the most famous sailing couple of the twentieth century. They lived on *Tzu Hang*, a forty-six-foot ketch from 1951 to 1969, and their adventures are legendary. Both Miles and Beryl wrote books. Two of Miles's best were *The Sea was our Village* and *Once was Enough*. I had the good fortune to meet both of them shortly before they died; that meeting will be described in a later volume of this memoir.

The first two parts of The Plan fell into place easily. We got through our year with less-than-outstanding marks, and we both found summer jobs. I don't remember where Don worked. I got a summer posting with the Department of Energy, Mines and Resources, whose Geographical Branch had an ongoing glaciological program on the Barnes Ice Cap bang in the middle of Baffin Island in the High Arctic. The job couldn't have fitted in better with The Plan. On an ice cap hundreds of kilometres from the nearest settlement, with food and accommodation provided, spending my earnings would be impossible. I expected the work to fulfill my concept of adventure more than adequately. I was not disappointed.

So it was that on May 11, 1967, I found myself on a Nordair flight departing Montreal at 2:30 a.m. (many hours late) en route to Frobisher Bay (Iqaluit) at the south end of Baffin Island in what were then called the Northwest Territories (now Nunavut). My only companion on the flight was John Clough, an American graduate student from the University of Wisconsin. A radio wave expert, he had already spent several field seasons in Antarctica. His job was to conduct experiments using a pioneering radio-echo sounding technique to measure the depth of ice to the underlying bedrock. I would start as his assistant for the short time he was taking part in the field program.

We spent an uncomfortable night on an elderly, four-engine, piston-driven DC-4. Having taken off in a heavy rainstorm, we were buffeted unmercifully by fierce headwinds, while the cabin alternated between freezing and sweltering. Four and a half hours later, we skidded down, shaken and grateful, at Fort Chimo (now Kuujjuaq) in northern Quebec about 650 kilometres south of Frobisher. The tiny settlement was cold and bleak. A few shacks huddled around an airstrip that was surrounded by stunted pines

Magic Travels

clinging to low, snow-covered hills. John and I wandered around for a couple of hours while the DC-4 was serviced and refuelled. Then we clambered back on board, frigid with cold, for an even bumpier ride to Frobisher Bay.

This outpost was dominated by the airstrip and a huge rectangular structure simply referred to as the Federal Building. This was where the government people stayed. John and I were given rooms and began to explore the facility. It was like a ship at sea. The inhabitants, once inside, seldom went outside again. The cafeteria was immense and forbidding. Above the serving counter hung a huge sign with what I read as a rather grim warning: TAKE WHAT YOU WANT. EAT WHAT YOU TAKE. It looked like both instructions were being followed obediently. Our fellow diners seemed uniformly obese. I imagined myself in the same condition by the end of the summer and shuddered. No doubt the sign was simply a means of discouraging waste, but for me, it took a lot of the pleasure out of eating.

First impressions of any new place often remain vivid forever. My introduction to the Arctic was no exception. It was impossible to call Frobisher beautiful. But the nearby Inuit village of Iqaluit (the name now used for the whole settlement) was worse. It is unacceptable now to use the term *Eskimo*, but at the time, we knew no better and used it without any concern for its colonial origins. Here I will speak only of the *Inuit* people.

The village of Iqaluit was awful. The ramshackle houses were little more than hovels, dirty, broken-down, and arranged chaotically with no apparent logic or concern for order. The rusty carcasses of skidoos and assorted parts, oil drums, general litter, and simple filth were all the more conspicuous against a backdrop of dirty snow, which was beginning to melt and reveal even more unsavoury detritus. Thawing dog excrement was everywhere. Huskies—chained to stakes, thin and mangy—howled and strained at their shackles, obviously eager to tear in pieces the two strangers invading their territory. These were not dogs you would ever want to touch, let alone pet.

I was overwhelmed with a sense of sadness and confusion. I wondered how I had grown up so completely ignorant of who the Inuit were and what kind of lives they were now living. How were such conditions tolerated? I remembered the superficial, patronizing treatment of Canadian Indigenous peoples in my high-school history books. We had been told they were of little consequence in Canadian history, that their culture was minimal, and their contribution to national life hardly worth thinking about. They were an inconvenient nuisance, as Thomas King brilliantly exposes in *The Inconvenient Indian,* his 2012 account of settler-Indigenous relations. My education, as it intended, had left me with the firm impression that they really were inconvenient and that their problems were innate and unsolvable, their own fault, not ours. As I walked around Iqaluit, I felt mortified by my ignorance and ashamed of what I was seeing. It was not the first time I'd had similar feelings. I had begun to feel uncomfortable about Canada's settler legacy when I worked with Dave in Jasper. But Iqaluit was a revelation. And not the last that I experienced in the North.

To make matters worse, the Inuit we passed on foot seemed indifferent, neither friendly nor hostile. Empty might be the best description. Not so the taxi driver who took us to Iqaluit. He made it abundantly clear he despised us, presumably because we were white. We were cheered up a little by the driver who took us the five kilometres back to Frobisher Bay where the white man's community, so clean and orderly, made a telling contrast with the squalor of Iqaluit. He might have been the Inuit equivalent of an Italian taxi driver. His hands scarcely touched the wheel, and we frequently missed roadside boulders or an adjacent ditch by a hair. He sang and laughed the whole way. As we approached one building, he shouted, "You want to go to post office?" Then, veering wildly, he declared, "You got no stamp!" and laughed uproariously. At another spot, "You want to go to telephone building?" Another violent swerve. "You got no dime!" And so it went. Every joke was followed by frenzied laughter. And yes, he was falling-down drunk.

∞

The next day, John and I met the incoming Nordair flight only to discover that the gear we were expecting was not on it. John decided to wait in Frobisher while I caught a DC-3 to a Distant Early Warning (DEW) Line site known as Fox-3 some 550 kilometres due north. I didn't know that this would be the first of many flights in the twin-engine, propeller-driven aircraft famous as the workhorse of the Arctic. It was recognized throughout the world as one of the most reliable aircraft for working in remote and difficult environments. The first one had been built in 1930, and the last, sixteen thousand planes later, had rolled off the assembly line in 1946. Many are still in service today. During the years I was to work in the High Arctic, I sometimes felt like I was living much of my life inside a DC-3.

But now, I was staring out at an unfolding landscape that was still as foreign to me as the far side of the moon. No trees were visible, not even shrubs, only immense expanses of wind-blown snow and rock. It looked utterly dead and frozen. Two hours later, we landed on a single airstrip that was hardly visible in the surrounding whiteness. A lone road curved up a gentle hill. Five kilometres away lay Fox-3. I would spend eleven days there before leaving for the Barnes Ice Cap, a further 110 kilometres north.

Fox-3 was one link in the DEW Line of radar installations that snaked across Alaska and far northern Canada. It was one of sixty-three stations designed to detect bombers on their way from the Soviet Union to drop nuclear bombs on US cities. Early detection would give the Americans time to counterattack—and make the destruction of the planet complete. Work on the DEW Line began in 1954 as the Cold War between the Soviet Union and the so-called Free World accelerated after the development of the first hydrogen bombs. Its construction was one of the most massive engineering and logistical feats ever accomplished. Like all the other stations, Fox-3 consisted of a number of prefabricated modules surrounded by antennas and plastic-covered domes. It housed about eighteen civilians and six military officers from both the United States and Canada. Monitoring the sensing equipment was continuous. Always men, they were isolated together for months at a time. Amenities included a bar, open twenty-four hours a day, movies, and copious amounts of food available around the clock. Much of the time, the weather precluded venturing outside for any reason. That was reserved for the catskinners who kept the road and airstrip functional.

Despite, or perhaps because of, the attractions of unlimited food, drink, and movies, the men on DEW Line duties developed a reputation for being strange. Excessive boozing probably didn't help. Arguments and fights were common. The isolation got to them, and the symptoms could be unpredictable. Connie, the station chief, exemplified what could happen. He was a small, fat, elderly man who was missing some of his fingers and some of his teeth. He chain-smoked large cigars and got tremendously anxious each time his leave (meaning that he had to *go south*) approached. For several years, Connie had found excuses to avoid who knows what dreadful fate he thought awaited him down south. He never spoke about it. Now he had run out of excuses and been ordered to take leave in no uncertain terms. He was not happy. The stress made him difficult to deal with, but he had the essential role of coordinating transport to and from the airstrip and arranging for a forklift to shift our supplies both on and off the planes.

To add to the stress, the members of our group arrived sporadically, and we were hardly welcome in the first place. We were outsiders from the South, and our job—to study an ice cap—was clearly silly, had no relevance to anything, and was obviously just another example of wasteful government stupidity. Above all, we were not staying long enough to earn any respect, while they were not getting the respect from us that was their due after already spending so many months, if not years, in this god-forsaken place.

Hostilities nearly always commenced at the bar. To be fair, the group I was in to begin with did not handle things well. We seemed to lack the ability to fit in. We were green and unaware of the stress we were inflicting on men who were deprived of ordinary social interaction. Dr. Peter Meyboom, a senior scientist who later became Deputy Minister of Fisheries and Oceans, was an exception. I am not sure what he was doing at Fox-3 as he did not take part in any of the future fieldwork. At any rate, one night he found himself in the bar when some of the station personnel were venting their frustration with our so-called scientific expedition. Gently, firmly, and with authority he took them on. It was a beautiful performance.

He explained in detail why this research was so important for Canada. He kept it practical, describing how the Barnes Ice Cap was the last vestige of the enormous ice sheet that had covered much of North America 40,000 years ago. This ancient chunk of ice now sat alone on the flat barrens of Baffin Island. It was 145 kilometres long and 60 kilometres wide, but it was not inert. It constantly shrank and expanded like a giant amoeba, always moving outwards from its middle to the edges. New ice formed whenever each winter's snowfall was greater than the amount lost during the summer melt. Then the ice cap would slowly advance over the land as it had done four times in Earth's recent history, when it had often reached far south of the Canada-US border. The opposite happened when less snow accumulated compared to the amount lost during the melt. Then the ice cap shrank.

Cores extracted from this ice can be used to document changes in the atmosphere's oxygen isotope ratios, which correlate with changes in global temperatures over time. Understanding past climate makes the future easier to predict. This information could, for example, enable viable future shipping through the Northwest Passage or predict ice conditions in the Gulf of St. Lawrence. It might prove vital to assess the likelihood of sea level rise or fall, which is largely dependent on how much of the planet's water remains frozen. The lecture worked, at least that evening. Peter had brought the temperature down, and the atmosphere became more tolerant, if not what could be called friendly. To help out, I finished with a magic show. After that I, at least, was in demand and we no longer had a problem getting help when we needed it.

For those eleven days at Fox-3, the weather thwarted our journey to the ice cap. Strong winds, blowing snow, and whiteouts hemmed us in day after day. Occasionally, for a change, everything disappeared into thick fog. On our second day at Fox-3, a C-46 aircraft crashed at a neighbouring DEW Line site, killing the two pilots and two passengers. The cause—whiteout conditions on top of poor flight planning. Any one of our party might have been on board. There was a marked decline in morale. Two days later, I watched anxiously as a DC-3 bringing in John Clough bobbed and weaved toward the airstrip. The crosswind was so strong that the plane landed on one wheel. The pilot headed straight to the bar, swearing he would never try that again. My roommate and fellow assistant, John Davis, not the most tactful individual, asked him why he had landed so dangerously and got an earful of colourful swear words from the pilot in return. It wasn't the first time, or the last, that I wanted to share with John the advice my father had given me before my first Jasper summer four years earlier: "Keep your eyes and ears open and your mouth shut." But John was a fourth-year student, and I was a sophomore, so I kept my mouth shut.

Like most southerners who arrived at the start of an Arctic summer, I had trouble getting used to twenty-four hours of daylight, but my acclimatization was made worse by Dr. Olaf Løken, our party chief. A glaciologist from Norway, he had ways of doing things that were not always, in my view, altogether logical, although I liked him personally. Olaf could not abide losing time to bad weather. For him, continuous daylight provided an opportunity to work whenever the weather permitted, regardless of the time of day. So it was that we would find ourselves down at the strip putting skis on the DC-3 at two in the morning or assembling newly arrived skidoos instead of eating supper. I could understand his reasoning up to a point, but over time, the lack of regularity in his regime became counterproductive. Exhaustion, irritability, and stress on normal bodily functions begin to take a toll. What's more, even in constant daylight, the sun gets close to the horizon at night, and temperatures drop. One night on the ice cap, after being woken from sleep for one of Olaf's unscheduled outdoor chores, my hands started to feel frostbitten. I promised myself that if I ever ran my own field party, work would start in the morning and finish in time to have a good supper and a proper night's sleep. If the weather happened to be foul all day, but then cleared up after supper, that was just too bad. The work would have to wait until the morning.

On May 23, Fox-3 was locked in by fog when we got up, but conditions cleared into a beautiful day by late afternoon. It had snowed, and the landscape looked magical in the sunlight. In high spirits, John Davis and I heard Olaf announce that we would be flown immediately to Generator Lake, the base camp at the south end of the ice cap. It would be our job to get it up and running. Simultaneously, others would be flown to another camp on the north shore of Inugsuin Fjord where the Geographical Branch had a semi-permanent centre of operations. Before we were halfway to the strip, a new fog bank rolled in. We decided to take a chance and stick to the plan. We rushed to load the DC-3 with the gear and supplies we would need, including four skidoos. Just as we finished, the fog miraculously dispersed. At the same moment, the truck radio announced that an Inuit had been shot north of Hall Beach. The plane was needed immediately. Cursing our luck, we unloaded as quickly as we could and watched our transport take off without us.

The good weather held into the next day, and the project helicopter arrived with David Harrison at the controls. I knew Dave by reputation as a revered member of the Baffin Island research team who had taken part in many previous operations. The helicopter (Dave got annoyed if he heard you call it a chopper) was a cut above the Bell model we had used in Operation Bow-Athabasca. It was a Jet Ranger, capable of carrying three passengers and the pilot and looked impressively modern and sleek. I soon discovered why Dave was so respected. A Brit with an MA in economics and geography, he worked as a high-school teacher during the winter but flying helicopters during the summer months was his true love. He was a force for sanity when it came to advising the scientists on the best way to accomplish the things they wanted to do. Everyone listened and did as instructed when the advice came from Dave. I came to admire him greatly and with his encouragement to imagine myself following a career path exactly like his.

On that day, I had little time to get to know him. Right after supper, the DC-3 would be returning to take John Davis and me to Generator Lake. As soon as we finished eating, we packed up the truck and headed to the airstrip. Olaf came with us. He had some last-minute advice. Here he was with two junior assistants who had never been to Generator Lake or anywhere in the Arctic before this. We had no knowledge of how to set up a radio, didn't have any idea what we would find at Generator Lake—what food might be available, what kind of tents we would be setting up, when the plane might bring reinforcement, etc. etc. What useful advice did Olaf have for us? Why, the most important thing we needed to know was the radio alphabet! "Alpha, Bravo, Charlie, Delta... Whiskey, X-ray Yankee, Zulu," I repeated back to him until he was satisfied that I knew it.

The pilot was in a hurry of course. The good weather could change at any moment, and we were already pushing our luck. We reloaded the four skidoos and everything else from the day before, only to learn that a low-pressure system was moving in. Generator Lake was only 100 kilometres away, so the flight was a short one. But with everything white and in a light that had already become flat as the storm approached, the ice cap and Generator Lake were barely distinguishable from the rest of the landscape. We made a dramatically bumpy landing on skis, bouncing off the tops of snow dunes that had been invisible from the air. Once safely on the ground, the pilot was desperate to get in the air again. The low-pressure system was close now, and he was unsure of the thickness of the lake ice now that some thawing had started. He also wanted his supper back at Fox-3.

He helped to throw our gear and the skidoos off the plane without even shutting down the engines, and within minutes he was airborne. Alone on the lake ice, we watched the plane disappear into the whiteness. A penetrating wind was already covering our mountain of gear in snow. The realization of our ignorance began to well and truly sink in.

"What do we do now?" John Davis asked.

Swallowing the exasperation I could feel welling up inside me, I answered carefully, "Let's get the gear and skidoos off the ice and get them up to the A-frame. A storm is coming, so we'd better get a camp set up as quickly as we can."

I had spotted the apex of a small A-frame sticking out of the snow a small distance away. It seemed reasonable to suppose it was not on the lake. "A-frame?" said John Davis. "What's that? Where is it?"

With a sinking heart, I gestured toward it. I no longer felt like John's junior.

"Oh," he said. "Do you think there is anything in it?"

I didn't answer. I was already assembling a Nansen sled and attaching it to one of the skidoos. "You start loading gear onto the sled," I told him. "See that rope over there? Use that to tie it down. I'll get the skidoos running and another sled. We need to hurry."

I got the skidoos started and attached a second sled. Grabbing a load of gear, I started arranging it on the sled, then looked up to find John Davis watching me. He hadn't moved an inch. His face registered a mix of curiosity and confusion about what I was doing. He genuinely had no idea what was expected of him or why. And he was meant to be my senior!

It wasn't the time to get angry. With what I imagined was infinite patience, I began to instruct John Davis on everything he needed to do. "Grab that brown case next to the aluminum poles and put it onto the sled. You see that other brown case over there? That goes on top of the first. Put them as far up to one end as possible, and that should leave room for those duffel bags. That's great, John!"

I did my own work as I guided him. Tying the gear down with rope was too much for him, but soon we had two sleds fully loaded. We drove them up to the A-frame. I went first to choose the route through the jumbled ice near the shoreline and to select a suitable site to unload the gear. Back and forth we went until every item was stored safely on shore.

It was now a kind of twilight and cold. The blowing snow stung our faces like needles. The A-frame was half buried in snow. Beside it was a tarp covering a cache of supplies that we were going to need. The cache was only visible here and there beneath the snow, and we soon found that it was encrusted in ice and could not be moved. Getting shovels and ice axes from our own supplies, I sent John Davis to the opposite side of the frozen cache. Despite the cold, I worked up a sweat hacking at the ice and shovelling it away. Unfortunately, I had neglected to give John specific instructions. I glanced up and found him watching me again with the same puzzled-fascinated expression. "What are we doing *now*?" it seemed to say. I was already getting the tarp free, so I told him (it wasn't a suggestion) to dig the snow away from the door of the A-frame and see if he could get it open. He did okay with that. It was just a bit of nuisance that in the process he piled the snow directly in the path between the door and the cache that we were trying to free. Together, we shovelled it to a new location.

Try as we might, we could not open the door to the A-frame. We hammered and hacked at the visible ice that seemed to be the problem but only made a mess of the door frame. Eventually we broke through the only window by prying around it with an ice-axe. Working our bodies headfirst through the opening, we discovered half a metre of ice holding the door. Had we stayed on the outside, we could have worked at it without success for the rest of summer. Hacking this ice out of the way, we finally got ourselves a usable door and were able to shift the ice and enough gear outside to be able to brew a cup of a tea. Upslope from the A-frame, we found a two-seater snow tractor (a J-5 Bombardier) together with a wanigan (a small hut mounted on a sled). Both were snow-covered, but inside the wanigan, we found sledge tents and foam mattresses. Rather than try setting up sleeping arrangements in the wanigan, which was tilted sharply to one side, we struggled to set up a tent. It was a design I was unfamiliar with, but it was extremely clever and ordinarily would have been easy to set up. Not quite so easy, however, for the first time and in a roaring gale.

Double-walled, the tent opened up like a large umbrella, its four sides held apart by poles that ran inside the two layers of fabric. The trick was to get each of the support poles planted into firm snow without any of the sides taking the full brunt of the wind. If the wind caught the tent before it was fully secured with all the guy ropes, it could easily become a sail and be blown away altogether. As usual, I needed to shout continual instructions at John Davis while working on my own side, but we managed to get it set up without too many missteps.

Pulling the foam mattresses in through the tunnel door of the tent, we assembled the sleeping bags (an inner and an outer) and were soon lying side by side in the semidarkness. The walls of the tent snapped like machine gun fire under the force of the wind while snow pelted against the fabric.

"Which sleeping bag do you zip up first, the inner or the outer?" asked John Davis.

"The inner," I replied. I heard the sound of a zip being pulled up, then after a lengthy pause: "How can I pull up the outer zip? I can't get my arms out of the inner bag."

I didn't answer. I would like to believe I really had already fallen asleep.

∞

Next, it was vital to get the radio up and working, but we couldn't do anything with it until the next day when we got a short mid-morning break from the continuous gale and snow that had lasted all night. There were no instructions. Nor could I be sure about how to orient the aerial. The radio alphabet was no help. I figured out how to assemble the rig and turn it on, but I had no luck raising Fox-3. I didn't even hear any static. The weather continued to be threatening, and the ceiling was low. There was no chance of a plane arriving that day, so we went back to pulling the frozen cache apart. I continued to instruct John Davis in simple, direct sentences. After a lot of hard work, we eventually recovered two skidoos, both wrecked, three sleds, a toboggan, a few cans of gas, and several boxes of assorted gear. As we were making lunch, the weather moved in. Further work was impossible. We retired to our sleeping bags and listened to a full gale pounding mixed rain and sleet against the tent. It was the only place to be warm and comfortable while waiting for better weather.

"Is this tent actually waterproof?" asked John. As I read in my diary fifty-four years later, it was "becoming increasingly difficult not to be sarcastic."

We emerged from our sleeping bags only to cook supper. The weather, with its accompanying noise, did not calm down until mid-morning the next day. We had to scoop snow from the tunnel entrance and burrow our way to the surface. After unsuccessfully trying the radio again, I tried to keep John Davis occupied as usefully as he was able. We sorted all the gear into various caches, including a large one that was composed entirely of food close to the front door. After cleaning out the A-frame, we completely reorganized the space, bringing in only the food supplies we needed immediately. As a result, we had places to sit and a more or less conveniently arranged kitchen area. The sloping roof precluded standing upright anywhere more than a foot or two from its centre line. The whole space would not accommodate more than four or five people easily.

In the afternoon the weather cleared up at last. CAVU (ceiling and visibility unlimited) was the acronym I was to learn later as part of radio speak. The sky was a brilliant, piercing blue. For the first time we could clearly see the rounded dome of the ice cap looming large above the jumble of moraines between us and its perimeter. Across the lake, sapphire-blue ice cliffs sparkled in the sunlight. A herd of caribou, their coats halfway through the seasonal process of turning from white to dark, wandered past the camp on the lake ice. What on earth could they find to eat? How could *anything* survive in this place? It was a mysterious world of stark beauty, and it left me breathless with excitement.

Just then we heard the hum of a plane approaching. A few minutes later it landed nearby, and we were greeted by the same pilot who had brought us. He was in a hurry again. We helped him dump a large number of forty-five-gallon drums of aviation fuel, naphtha, and gasoline onto the ice. Then off he went without another word. It seemed prudent to get the drums off the lake quickly. With the sun shining, the new snow was already becoming slushy, which made the skidoos slower and less manoeuvrable. But quick was not in my companion's vocabulary. He reminded me of Ollie, the horse in Jasper who had to be guided around every bend in the trail to prevent him from proceeding straight into the bush until he hit a tree.

Even without the extra burden of John's gormlessness, it was difficult work. Each drum, weighing about 175 kilograms, had to be manhandled onto the Nansen sled and held in place (by John, very tenuously) while I drove the skidoo to our chosen site well clear of the A-frame. It wasn't long before John complained that it was too difficult to hold the drum on the sled. We reversed roles. Now he seemed incapable of driving the skidoo without messing up in the jumbled ice near the shoreline. In the end, I told him to wait on the ice with the drums and just help me load each one onto the sled. Then, as he helped by watching me, I would tie the drum on using my cowboy packing techniques and drive it alone to the cache site. There I could roll the drum off by myself before rejoining John for another load. There must have been around twenty-five drums in all. I was exhausted when, at 10:00 p.m., the last one was safely on land. This coincided with the arrival of another load of drums.

But now, there were more than just drums. Two people got off the plane with them. I had not met them before. John Richardson and John England were their names. As they got their bearings, I could see John Richardson talking incessantly to a blank faced John England. As they drew nearer I could hear.

"What's that over there, John? Is that the ice cap? I've never seen an ice cap before. Are we going to be going up there? Where are we going to sleep tonight? Are there polar bears in this area?"

I couldn't believe my ears. How could there be two John Davises on the planet, let alone both together on Generator Lake at 10:00 p.m. on May 26, 1967? My eyes met John England's. Words were unnecessary. We understood each other. It was telepathic communication at its most profound. Transmitted by sympathy. I knew exactly how he was feeling, and he knew I had been suffering in exactly the same way. But there was also relief in his eyes, and I felt the same. We were no longer alone with only an alien for companionship. Together, we would survive this ordeal. That night John England and I shared a tent. He had a bottle of whisky. We drank it. Our laughter and our jokes, our tales of what we had been through over the past week, rang loud and clear for all to hear. We didn't care. We were already friends.

It was another nine days before our party was organized enough to begin the work we had come to do. The dreadful weather was relentless, and our preparatory work could only be done unevenly and sporadically. But the party continued to grow in numbers, and more provisions, fuel, and gear were dumped out of the DC-3 whenever there was a respite in the weather. On the fifth day since arriving at Generator Lake, I finally got some joy from the radio, and we were able to talk with Fox-3 and the Inugsuin base camp. I spent a lot of time on my skidoo, marking out an airstrip on the ice using flags and smoothing out the snow dunes. There was no doubt about it. Skidoos were fun. John England and I found racing them irresistible. The frequent whiteout conditions made our antics even more challenging. And thrilling.

In a whiteout, all potential obstacles, such as snow dunes and ice chunks, become invisible. Even the skidoo's tracks and the footprints you have just made in the snow are impossible to see. It is eerie and otherworldly. With everything white and no visible horizon, it is like being suspended in emptiness. It is not unusual to lose your balance and keel over into invisible snow. It is perfectly possible to get off your skidoo thinking it has stopped, only to discover, as you are bowled over into the snow, that it has not. Perspective is always uncertain. More than once, I thought I was seeing the dark form of a caribou in the distance only to run over a lemming crossing right in front of the skidoo. It is not uncommon to become airborne when going over a dune too fast but not know this until you feel the jolt of landing. Spills happened frequently, and one of our party injured his knee quite badly.

We took the small windows of good weather to explore the area. Going northwards took us through the moraines that lined the edge of the ice cap. To get to the ice cap itself, a route had to be found through a jumble of steep, chaotic mounds and ridges of rock rubble. These were glacial deposits composed of till, a mixture of all sediment sizes ranging from large boulders to the finest clay. Once through this obstacle course, the smooth snow-covered ice sloped upwards to end at the horizon. Only the presence of the horizon, when it could be seen, provided any sense of orientation, but it disappeared whenever cloud cover merged the horizon with the sky. The result was a whiteout. And they were common. It didn't take exceptionally bad weather to cause a whiteout. All you needed was to lose the horizon.

Once in a whiteout, you were for all practical purposes lost in space. Global positioning had not yet been invented. And the ice cap, perversely, was situated over a gigantic deposit of iron ore that made a compass needle spin or point in any direction at random. That left only two possible indicators to maintain a direction, both temporary, both uncertain. The first required a wind to be blowing. Then you could, for example, drive your skidoo with the wind hitting your left cheek or some other part of your face and body. This could work quite well but only if you knew your direction before the whiteout hit. The same reasoning could be applied to the orientation of the ripples and dunes formed in the snow, although in a whiteout they were invisible, and you had to feel them through the motions of the skidoo. These indicators became important for me. I had a tendency in a whiteout to think that I was constantly bearing left (or right). If I gave into my directional delusion and attempted to adjust my course accordingly, I drove in circles. (Years later, I suffered from the same directional delusion when sailing in fog, even to the point of doubting my compass.)

Apart from that peculiar tendency, my navigational instincts became quite good. I often preferred to be alone or to take the lead rather than follow someone who didn't understand such niceties. Driving the ice cap had its dangers. While most of its margin graded smoothly into the rubble of moraines, where a lake was present, there was usually a vertical cliff. On one occasion, while driving back to the Generator Lake base camp in whiteout conditions, I failed to notice that I had swung too far west. I not only missed the moraines; I nearly drove off the cliff edge onto the ice cover of Generator Lake. Only a glimpse of bright blue ice alerted me to the peril. I flung myself off the skidoo, which somehow stopped just before going over the edge. The loaded sled I was towing, however, halted only when it rammed me hard in the chest.

One fact about ice cap life, apparent to nearly all of us, was the importance of sundogs and halos. Caused by the refraction of sunlight passing through ice crystals in the upper atmosphere, a bright spot to one side of the sun (a sundog) was often seen. Another spot could form symmetrically on the opposite side. At times, as many as four spots might surround the sun. They were first described, as far as we know, by the ancient Greeks, who thought they foretold bad weather. On the ice cap, we confirmed that the Greeks had got it right. We observed that high winds and whiteout conditions would reliably follow within hours of sighting sundogs, and the severity of the storm increased with the number of sunspots observed. The worst weather followed a halo when the four sunspots coalesced into a single ring around the sun. If we were working far from camp, we knew to head back quickly while it was still possible to follow our skidoo tracks.

Without John England, life on the ice cap might have been intolerable. His enthusiasm and sheer joy at being in the Arctic was contagious. We worked together as often as possible, which meant we took on jobs that we were pretty sure would stymie our colleagues. It was a triumph when we greased and tinkered with the J-5 Bombardier and eventually got it to go. The wanigan was attached to it and the insides cleaned up to serve as a mobile headquarters and dining area as we travelled the ice cap. There was only one snag. The J-5 was steered by pulling on the left or right brake levers to stop one or other of the treads, but this one had a glitch in its left brake and could only turn right. On the dead flat portions of the ice cap, this wasn't an insurmountable problem, but going up the steep side slopes was a nightmare. The moment the Bombardier went a little too far to the right, it became impossible to turn it back upslope. With the wanigan in tow, the J-5 wasn't powerful enough to go directly upslope, but elevation could be achieved in a series of switchbacks. To make the turn to the left required the passenger to jump out and push the cab around to the other tack without stopping the vehicle. It didn't take long to become exhausted. The problem was finally fixed with the help of our helicopter mechanic.

Our largest and most powerful skidoo was the Road Master. Its job was to pull a decrepit Nansen sled modified to carry John Clough's radio-echo sounding gear. Why such an old sled was chosen for such an important job, I do not know. It gave us nothing but trouble. John Clough and I assembled the rig by first installing plywood sidewalls to the sled, then using a tarp for a ceiling. This made the interior dark enough to see the oscilloscope and other electronic dials. Foam mattresses were laid down so that John could lie inside and adjust the radio signals. These penetrated the glacial ice, bounced back from the bedrock, and were received by aerial receptors. Analysis of the return times revealed the thickness of the ice.

The gear and batteries required to run the experiment were heavy and made the sled unwieldly. This first became apparent when we attempted to get it all up onto the ice cap and start John Clough's research. The great day finally arrived on Sunday, June 4. The day before, we had managed to get the J-5 up onto the ice cap. This seemed to call for a celebration that evening, which proved to be a misjudgement. Three of us (John England, John Clough, and I) ended up carousing our way through a substantial amount of whisky and didn't get to bed until sometime after 3:00 a.m. We were a bleary lot when we started our day four hours later, but at least it was warm and sunny. Despite our hangovers it was impossible not to be in good spirits.

It was a hard slog to get the Road Master and the sled of radio sounding gear all the way up to the first line of survey stakes. Labelled the K-line, this was one of many survey lines criss-crossing the ice cap. They were identified by stakes drilled into the ice about a mile apart. These had been installed over the years to track snow melt and accumulation. The sunshine, though pleasant to work in, had softened the snow causing the sled to bog down frequently and the Road Master to spin its tread until it too was stuck. After much pushing and floundering about in deep wet snow, John Clough and I often collapsed in exhaustion and had to rest up before the next effort. It was always at such moments that Olaf chose to drive up on his skidoo. I began to notice that he never actually did much to help. Instead, he appeared to drive great distances between the working parties only to give orders (often counterproductive) or advice (generally futile) on how to do our jobs. I remember that his mitts had distinctive fluorescent red backs that he seemed to think gave him an air of authority.

"Perhaps you should try to keep moving rather than constantly taking rest stops," he advised us when he found us taking one of our much-needed breaks.

We just stared at him. He must have got the message as he only repeated himself two more times before he took off, no doubt to offer more helpful advice to the other party, which was drilling in new stakes on the K-line. Olaf occasionally had difficulty keeping up the morale of his team.

We finally made it to the K-line, where John's radio sounding work was to begin. Because the technology was still new and largely untested, it took him some time to get any meaningful measurements. He discovered that the best results were obtained when the two aerials (sending and receiving) were located as far apart from each other as possible. The aerials were held horizontally on two long poles. One extended out on both sides of the Nansen sled. The second, to maximize the distance between them, was mounted on the Road Master and separated from the sled by a long tow rope. The complete set-up looked extraordinary; the aerials made the sled and the skidoo look as though they had sprouted wings.

Our procession consisted of me driving the Road Master down the K-line, John Clough lying inside the sled working the equipment, and John England standing on the back of the sled as though on a caboose, shouting out the distance every 200 metres as registered on an odometer, the last towed item of the train. Up and running, it was an impressive sight, and Doug Hodgson, one of our colleagues in the party, memorialized the occasion by taking a number of pictures. Unfortunately, one of our aerial wings smashed the windshield of his skidoo. The radio-echo profiling continued throughout the afternoon and evening until finally the weather moved in again. Leaving the equipment at one of the K-line stakes, we returned without mishap to Generator Lake, having completed our first day of productive work on the Barnes Ice Cap a mere twenty-four days after leaving Montreal.

∞

The day after our first foray onto the ice cap, we faced an even bigger challenge. To start the next phase of the work we had to move from Generator Lake and set up camp on the I-line about forty kilometres north. It was another fine day, but as we started packing up camp after an early breakfast, I glanced up at the sun. There was a spectacular halo around it. I pointed it out to John England. He called out to Olaf.

"Olaf, there's a halo around the sun. It could get bad. Do you think we should go?"

"No, no," replied Olaf, apparently meaning, "yes, yes. It should be okay." There was no point discussing the subject further.

As if on cue, four hours later the wind started up and visibility disappeared just as the last sled had been packed and the last knot tied. Off we went anyway. What followed could have made a comedy movie except, of course, that nothing would have been seen on the screen except white. And it wasn't all that funny either, not at the time at least. To start with, everyone took off in different directions and immediately got lost. All they could then do was wait for someone to find them again. The J-5, pulling a heavy, overloaded sled, kept getting stuck among the moraines, and many hours were spent pushing and heaving. After painfully switch backing up the slope of the ice cap, the Bombardier and wanigan finally made it to the top. There, we assembled as a group. Olaf, gesticulating with his fluorescent, red-backed mitts, instructed everyone to follow him. By then the wind was howling, and a glaze of frozen sleet coated everything. We were going north to the I-line come hell or high water!

There were only two small questions. How was Olaf going to know the way? And how would he know when we had arrived? We knew better than to ask. In a single-file armada of skidoos, sleds, and the flagship of the fleet, the J-5 towing the wanigan, we sallied forth. Goggles were essential to protect our eyes, but the freezing rain made them useless for seeing through. The best you could do was to follow the indistinct dark shape in front of you and hope for the best. Occasionally Olaf would circle around his wagon train to check on everyone and fill up with more gas if necessary. More often, he pulled too far ahead and had no idea when a skidoo broke down or a sled had tipped over, not to mention the delays caused by the temperamental J-5. If and when he did look back and couldn't see his caravan, he would get lost trying to find us. The hours ticked by. It was miserably cold. Our mittened hands, painfully struggling to keep their grip on the steering handles, felt it most. I had the impression from my wind-direction navigation that we were not keeping a very regular course. We had been up for about twenty hours, travelling for about sixteen of them, when Olaf, to the relief of all, halted the entourage and announced, "We have arrived. Let's set up the camp."

No one needed a second invitation. Tents were hurriedly erected, a makeshift meal was consumed inside the wanigan and, with groans and sighs all round, we retired to the warmth and comfort of our sleeping bags. This is how I describe the following hours in my diary:

I dimly remember hearing the wind howling and the snow blowing, and when I woke up, I couldn't believe that it was 8:00 p.m.—about fifteen hours sleep. I must have been tired! There was nothing to do but stay in the tents. At midnight, Olaf called us out for food, and outside, it was impossible to see anything. From the tent, it was barely possible to make out the wanigan for the blowing snow. After eating, we talked a while before struggling back to the tents. Stayed in the sack for another fifteen hours but got up when I could stand it no longer. The wind had stopped, but it was still foggy. The rest soon got up, and we spent an exhausting time digging out all the tents, sleds, skidoos, and even the wanigan to clear the snow drifts.

Finally, after almost two days huddling in our tents, the weather cleared enough for me to take my skidoo on a reconnaissance mission. It didn't take long. I was soon stopped by the cliff edge that looked out over Generator Lake. I stared across at the A-frame. We had moved a net distance of about five kilometres.

∞

Our botched entry onto the ice cap was merely a taste of how the rest of the month would go. The work consisted mainly of finding the different lines of stakes that were scattered from one end of the ice cap to the other. These were simply long aluminum poles that stuck out of the ice, but they were exceedingly difficult to find. They had been set too far apart from each other to enable the next stake to be seen from the one you were at. With frequent whiteouts, finding the next stake in the line was largely guesswork. It could take a long time, and there was plenty of opportunity to get confused.

To give you an idea, John England and I completed our work at one stake in a whiteout and then tried to set a course for the next one. When too much driving time elapsed without any sight of our goal, we had to acknowledge that we must have missed it. Turning the skidoos around, we tried to follow our barely visible tracks to get back to the stake we had come from, reorient ourselves, and make another try. But no matter how slowly we drove and how carefully we peered down at the snow, we finally realized that the markings we thought we were following were phantoms. We really couldn't see the tracks at all. Pulling to a stop, we sat on our skidoos and looked at each other, neither of us wanting to acknowledge that we were lost - again. Suddenly, looking beyond John's shoulder, there the stake was, just in front of us. Congratulating ourselves on our good fortune, if not our skill, we sped toward it. There we celebrated with a square of Baker's semi-sweet dark chocolate, which had become one of our staple foods.

"John," I asked gently, "how come I don't see any yellow pee marks in the snow? I'm almost sure we both took a leak when we were here. In fact, I can't see any evidence that we were ever here at all." We both began to laugh. As understanding sunk in of what we had done, we laughed even harder. The label on the stake we were now parked beside revealed that this stake was the one we had been looking for in the first place.

The work down each stake line was difficult and could go wrong in any number of ways. The worst part was adding new stakes at intervals between the old ones. Using a two-person, gas-powered auger, we used interlocking drill bits. Each bit could drill the hole down about three feet, after which the apparatus was hauled out and another bit added. With increasing numbers of bits, the auger became heavier and heavier and the work harder and harder on our backs. The auger engine blew exhaust directly into the face and was piercingly loud. We were all somewhat deaf by the end of a day (or night). Ear protectors were either unknown at the time or considered unnecessary. I know I never thought of them, something I now have reason to regret. The augers, which were old, needed constant attention and tinkering to keep them running. The drill bits frequently didn't fit together well and often jammed in the hole, stalling the motor. Our strongest efforts, assisted by even stronger cursing, weren't always able to rescue the stuck bit. There was little to be done but lose some of the bits to the hole and start a new one.

We also dug pits down to the snow-ice interface and made notes on the stratigraphic succession and measurements of the thickness and density of each observable unit. Typical sequences were surficial snow at the top followed by increasingly denser and more crystalline snow until the fern layer (the recrystallized granular snow intermediate between snow and glacial ice) was reached. These data would later be compared with observations from previous years. The idea was to assess the mass-balance of the ice cap in order to answer one burning question: was the glacier building or wasting away?

The weather, as you will have guessed by now, was seldom kind. Being lost in whiteouts was almost the norm. When not lost, we had to spend so many hours inside our sleeping bags listening to the wind howl and the sleet pound the sides of the tent that we sometimes felt we would go mad. We were frequently three to a tent, which left no room for anything but bodies lying like mummies beside each other. Calls of nature were put off for as long as possible. When the strain became too much, you had to manoeuvre yourself out of the double sleeping bags, grope around for sufficient clothing, go into contortions to get your mukluks on, and finally crawl out the tunnel exit into the maelstrom outside. The return, despite the inevitable snow carried in on clothing, was joyous, made more so by knowing that the others still had to face the same ordeal. I had brought books with me—even a Bible, which I had thought might be a good thing to read at the top of the world. I got only a few chapters into it before moving onto Solzhenitsyn, where I found, unlike the Bible, wisdom that has remained with me to this day.

The irregular routines inflicted on us by the weather made meals sporadic. Our food supply was mostly freeze-dried in cans. Weighing only a few ounces each, the cans were great for helicopter transport, but the food they contained was less than great eating. Meat was limited to freeze-dried pork chops, mini-steaks, and *Burgy Bits*, the brand name for little chunks of ground beef that bore a close resemblance to dog food. "Soak for one hour in water and cook as normal," read the instructions. This was an outright lie. One hour of soaking did nothing at all. Eating a fried steak was like chewing cardboard and tasted worse. The pork chops were more edible, but only if you could soak them for twenty-four hours, not an easy thing to do in the conditions we faced. On an ice cap just making water takes a lot of effort. About ten litres of snow is needed to get one of water. It helped if the soaking was done in warm, or better still, hot water. To ensure a ready supply, I wedged a covered saucepan containing a supply of the chops on top of my skidoo's exhaust manifold when we were travelling. This rehydrated the chops perfectly. Unfortunately, they were only edible if you enjoyed the taste of exhaust fumes.

In more ambitious moments, I would sometimes go to great lengths to make spaghetti and pasta sauce. Using freeze-dried onions and mushrooms together with some Burgy Bits, plenty of soaking, plenty of simmering, and as many spices as I could scrounge from the supplies, it was possible to make an acceptable Bolognese. On June 21, for my twentieth birthday, I decided to treat myself and John England to this feast. By this point, we had reached the far north end of the ice cap. Crouched inside the sledge tent with a large pot simmering on a Coleman stove, I knew our supplies were low and that we would soon run out of naphtha. John had already radioed Olaf about this shortage. Our chief had advised that gasoline, which we still had plenty of, would probably work just fine for the stove. John had pointed out that using gasoline inside a tent could be dangerous and the Colman stove might explode.

"You have an extra tent with you," Olaf replied. "A single tent is expendable."

"Actually, Olaf, I was more worried about the people in the tent," John said. The conversation was abruptly cut short.

As I stirred my gourmet delight, the Coleman stove gave a last flicker and went out. The sauce had been coming along nicely but was nowhere near ready. As I was pondering the matter, John crawled into the tent. The snow beneath it had become soft and unstable with the onset of warmer weather and, without warning, the floor shifted. My full kettle of pasta sauce fell off the stove, spreading in all directions, soaking everything in its path and pooling in the depressions. I was more than a little depressed myself as I mopped up my *spécialité de la maison*, which now looked more like vomit.

Olaf and his assistant left the ice cap before us. He wanted to get south, and who could blame him? It wasn't an easy escape. My diary recorded some of the details.

They left late in the day and the weather was threatening. We heard later that it took them about four days of travelling in whiteouts and camping in stormy weather. They missed Generator Lake by a considerable margin, and Olaf was very fortunate to catch a lateral (a connecting flight between Arctic settlements) *to Hall Beach.*

I stayed behind with a skeleton staff to complete the small amount of remaining work. This is how my diary summed up these last few days.

We set up our final camp at C-29 (the twenty-ninth stake on C-line) *and did work on the D-line which branched off to the west from C-22. This work was done only after another forty-eight hours in the sack. At D-3, the last stake we had to drill, the auger stopped working. Thoroughly discouraged by the length of time doing nothing, eating cold meals (meals?), and living like pigs, we packed up early on the morning of the 25th and headed for Generator Lake.*

We got away by 8:30 a.m. and the weather was CAVU. We did measurements along the C-line up to C-44 and dug a couple of pits. At 12:30 we were on our way with two skidoos that were very, very shaky and two immensely laden sleds. Our speed was very, very slow. The hours ticked by and soon, just for a change, the weather moved in causing a whiteout. We hit H-9 so we knew we were too far east. In the whiteout we managed to find I-19, and we drove up the line a little way looking for the Bombardier tracks but didn't find them. We were, for a while, going to set up the tent and wait for the whiteout to lift, but I felt I could get enough of an orientation from I-20 and by keeping the wind on my left cheek to stay in the right direction. Wonder of wonders, the weather cleared and we hit Generator Lake dead on. While driving off the ice and through the moraines, I caused a temporary delay when my sled tipped over, but we were at the A-frame by 10:00 p.m. My back was exceedingly sore which, on reflection, must have been caused, not by the skidoo driving, but when I righted the heavy sled. Our greetings and cheering were loud and spontaneous, and after a large hot meal of fresh T-bone steak, we all felt great.

∞

When I first met John England, I was particularly taken with a pair of his boots. Across the backs of them were printed the words "Brutally Strong." For some reason, I thought this was funny, perhaps because I associated *brutal* with something violent and unpleasant. Others must have felt the same way. *Brutal* began to be used ironically by the whole team whenever anything enjoyable happened. "Man, that was brutal!" if we managed to drill a hole without mishap, for example, or after an especially good cup of coffee. It was usually spoken almost as a sigh, in a tone of wonder. It got to be so commonly used, and just about always brought a smile, that it almost replaced the ubiquitous *fuck*. And that was what the days were like following my departure from the ice cap. Absolutely BRUTAL!

To start with, I was left on my own at Generator Lake. Dave, the pilot, arrived on the first morning to ferry all the others to the Inugsuin base camp. I was left to man the radio. He had brought some fresh eggs, so I was able to enjoy a cheese omelette for lunch. After much trial and error, I managed to get a kerosene heater going and spent a comfortable interval reading before cooking supper and going to bed. I didn't even get lost on my way to the tent.

In the ordinary state of affairs in a person's life, one is seldom alone. Lonely, yes, but it is rare to be absolutely alone with not another person even remotely close. It was an experience I was to repeat in later Arctic seasons, and I am grateful for it. It was difficult at first. I worried about how to pass the time and made myself do jobs that I fancied needed doing. I waited impatiently for the scheduled radio skids and the chance to talk to someone. But then, after a day or two, I began to resent these interruptions of my solitude. They felt like intrusions.

As the days passed, I found the isolation more and more satisfying. Once a group of wolves visited. And with some of the landscape now free of snow, odours and colours began to emerge. After weeks of frozen whiteness, such changes were delicious. All sorts of beautiful flowers, Arctic poppies, saxifrage, and pink clusters of wintergreen appeared almost overnight, sometimes even pushing through the melting snow. The heavenly aroma of moist soil made me realize how deprived the senses can become if they have nothing to do. I was thrilled even by the tinkling sounds of the small streams that seemed to be everywhere. It was spring. The land's awakening was like a rebirth. It was intensely emotional. I felt reborn myself, exultant to be part of it. On a more practical level, I no longer had to melt snow to get water. I spent most of one afternoon heating water from a small brook and luxuriating in the sensation of washing my body for the first time in over a month.

The euphoria, if anything, increased when, one midnight, Dave arrived to fly me to the Inugsuin base camp, seventy-five kilometres to the east. From the barren monotony of the ice cap, we flew to the fjords of eastern Baffin Island, possibly the most beautiful scenery in the world. To the north, the sky was a fiery red. Ahead, I could see the beginning of the line of fjords that extended another one hundred kilometres to Baffin Bay. Their sides rose a thousand metres from the water. Snow and ice fields capped the mountains between them. Long arms of ice extended down sidewall valleys. Rivers were flowing, and waterfalls were plentiful.

After the ice cap, it was delightful to arrive at the Inugsuin base camp. Perhaps I'd had my fill of soul-restoring solitude by then. I was led into a large, cheerful building that housed the kitchen area, full of tables and chairs, and then to an annex with bunk beds and windows that overlooked the fjord and provided an unsurpassed view that never failed to take my breath away. There seemed to be no immediate plans for work, and I found myself with little to do for those first two weeks of July. I took long exploratory walks and climbed one of the nearby mountains. I busied myself cooking, making bread and pies now that I had a propane stove to work with. I constructed a plywood roof on the aging supply tent.

One morning I awoke to the sound of Inuit voices. A small hunting party composed of two brothers, Butch and Salomonee, and one of their friends had arrived from Clyde River. In future seasons, Butch and Salomonee were hired to work with us on the ice cap, and I got to know them well. Butch, the youngest and the same age as me, was an extrovert and prankster. He spoke the best English, and, since it was a wet day, he decided to teach me the Inuit symbols for the alphabet instead of going hunting. They fill a page in my diary. In return, I did some magic and came to a conclusion confirmed by all my future interactions with Inuit. They make the best audience for a magician in the world. As Butch and Salomonee watched me, pure rapture registered on their faces. Their laughter and acceptance of what they had seen was absolute. There was not even the slightest curiosity about how a trick was done, only mystified joy.

The following year, I used the stage name *Anko* for my nightclub shows in Australia. Chosen for me by Butch, it means shaman in Inuit. This Inuit connection went over well down under. I learned quickly that Aussies are drawn to anything salacious. All they knew about the Inuit was that they shared their wives. This appeared to give rise to glorious and wistful fantasies, at least among the men, but I doubt if their daydreams would have survived any actual knowledge of the complexities and dangerous pitfalls of traditional Inuit polyandry. Such as murder. In any case, I didn't spoil their fun by telling them that the arrival of Christian missionaries had long since put a stop to the practice.

At the end of that afternoon at Inugsuin, the weather cleared up and the Inuit hunting party disappeared to look for caribou. They returned at 4:00 a.m. empty-handed. Later that day, as they got ready to return to Clyde River, we saw that the ice had drifted a couple of kilometres down the fjord. They were now stranded at our camp, unable to reach the ice with their skidoo. In any case, their skidoo was out of commission and couldn't be driven anyway. The ski had broken. While one of our party repaired it with welding equipment, I rigged up a Nansen sled with an empty forty-five-gallon drum lashed to each side. The flotation would be sufficient to bear the weight of their skidoo and sled. Using our Boston Whaler with its ten-horsepower motor to tow the Nansen and its load, I ferried the hunters to the ice edge. Even to my inexperienced eyes, it looked like sufficient melt had taken place to make the ice dangerously rotten. Large, shallow meltwater lakes overlay the ice everywhere.

Butch and his companions did not seem in the least concerned. I marvelled at their nonchalance and wondered what they would have done if we hadn't fixed their ski or not been able to get them to the ice. What food did they have for the seventy-five mile journey to Clyde River? None as far as I could see. Where and how were they going to sleep? Who would know if their skidoo gave up the ghost, or worse, if they went through the ice? As we stood on the thinning ice surface, with the sun low in the north, their short statures were silhouetted against the sky. It was a romantic image that has never left me. Salomonee spotted a seal. We watched in silence as he stalked it across the ice. He missed, and it disappeared down its hole.

Their insouciance seemed bizarre to me, but it was consistent with their overall deportment. They had arrived out of the blue with no concern for the fact that everyone was asleep at the time. They had made themselves at home, eaten our food, come and gone as they pleased, had their skidoo fixed, and got a lift to the ice to make their homeward journey possible. Yet there had been not a single sign or gesture of thanks from them. Even considering the language difficulties, it did not seem right to me. A day or two later, I read Doug Wilkinson's *Land of the Long Day* and began to understand. Their traditional way of life demanded mutual sharing of everything. It was a matter of survival. Other people's hospitality had always been taken for granted, and still was. Gratitude, at least in the way we understood it with all its accompanying vocabulary, was a foreign concept. Knowing and accepting this made my future encounters with the Inuit much easier and more enjoyable.

∞

I was finally on my way to work again with John England. Pleasant though the interlude at Inugsuin base camp had been, I was keen for a change of scene, which started with one of the most beautiful helicopter flights I have ever experienced. Alone with Dave, we departed Inugsuin near midnight en route to Fox-Charlie (or Fox-C), an abandoned DEW Line site on the south side of Ekalugad Fjord. I had seen enough of a Baffin Island summer by now to understand that the weather could be amazingly variable. Some days were so hot and sunny that the air became thick with enormous mosquitoes. Fortunately, their size made them ungainly. They flew slowly and were easy to swat. It sometimes got hot enough to make a plunge into an icy lake a glorious relief. But good weather never lasted long. More often than not, rain and mist arrived in short order, obscuring the landscape and hiding its stark beauty. When the wind blew, it became intensely cold.

But that night was magic. Flying the 110 kilometres southeast to our destination, we crossed over one fjord after another. The sun was low on our left. In perfect balance with it on our right was a full moon. Both were almost supernaturally large, and they seemed to hang immobile in the sky. The deep orange orb of the sun competed with the white orb of the moon to cast rich dark shades of blue across an infinite, cloudless sky. Below, the snow on the mountain ice caps sparkled. Long tongues of valley glaciers spilled down the steep slopes into the fjords. Together, we sat in the tiny helicopter in silent communion with the majesty of the scene. Even the noise of the helicopter seemed dwarfed, overwhelmed by the magnitude of the beauty around us. Like the sun and moon, we were also suspended in time and space. Finally, I leaned across to Dave and said, "Just drop me off at that spot right there with a pair of skis. That would be the run of the century." He just nodded and smiled.

Fox-C had been deserted for four years. The site itself was on top of a nearby mountain, but our camp was on the beach where supplies for the DEW Line site had been brought in by sea. It was impossible not to be depressed at the mess left behind when the station was abandoned. Several ruined trucks and hundreds of empty fuel drums were scattered throughout the area. Two huge rusting gas tanks dominated the site, and our Parcol, a Quonset-like structure made of fabric, had been set up between them. I decided that the iron-rung ladders to their tops would make good escape routes in the event of polar bears, but we never had to put them to the test.

The shambles around Fox-C were the first of many insights into the military mind that I acquired over my years of working in environmental geology. I came to understand that it consistently operates with no regard for the consequences of the activities it dreams up. It took until 1989 for the Department of National Defence to begin the long-overdue cleanup of DEW Line sites. Predictably, it turned into the single most expensive environmental project (at that date) ever undertaken. When the cleanup was announced as finished in 2014, it had cost $575 million. The remediation work included demolishing infrastructure, removing contaminated soils, stabilizing landfill sites, and constructing new landfills that, we were told, *will stand the test of time.* What took three years to build required twenty-five years to clean up, and a federal-government monitoring program has been promised to ensure the sites remain safe for at least the next twenty-five.

John England and I did not remain long at Fox-C. We now had an aluminum, five-metre Boston Whaler with an eighteen horsepower outboard. Our job was to explore Ekalugad Fjord for raised beaches and marine limits. Fjords are classic glacial landforms that were caused by ice scouring the bedrock to form the characteristic U-shaped valleys seen in many glaciated regions of the world. At the height of the glacial episodes that advanced and retreated across various landmasses, the sheer weight of ice (three kilometres thick) depressed the earth's crust. In addition, water that evaporated from the sea and produced the snow to grow the ice caps caused sea levels to lower about 125 metres around all coasts of the planet. As the ice melted, returning water back to the sea, the loss of weight enabled the crust to bounce back. Of course, the return of the water to the sea created a rise in sea level, but the speed of the crustal emergence was even faster. The evidence for this can be seen in sequential strand lines (raised beaches) that formed as the land emerged from the sea. The first beach to form is the oldest, and after the emergence of the land, it becomes the highest beach in the sequence. Its elevation is known as the marine limit. Each beach downslope from the marine limit is progressively younger all the way to the active beach at the present shoreline.

Our task was to collect data to work out the rate of emergence (the amount of crustal uplift over time) that had occurred throughout Ekalugad Fjord since deglaciation of the area. To make such a calculation required two types of data, the first being an accurate elevation for as many raised beaches as we could find, the more in any one sequence the better. Our tools were a survey rod (a flat-sided, collapsible pole with clearly marked graduations from zero at the bottom to three metres at the top) and an optical level (a theodolite). After setting up the theodolite on the active beach, the *rodman* marches toward the first raised beach, being careful not to go higher than the three-metre height of the rod. Holding the rod vertically, the surveyor uses the theodolite to take a measure of the height of the land at the rod. With the rodman remaining in position, the surveyor then carries his equipment further inland to measure the height of his new position relative to the rod. In this way, surveyor and rodman leapfrog each other up to the marine limit, taking a measurement at each raised beach. We developed a set of hand signals to indicate when a measurement was finished or that the rod needed to be straighter. Distances could be too great for shouting. Wind and rain slowed us up considerably, and miscommunicated signals, particularly when the weather was bad, could lead to alarming irritability.

Once the marine limits had been ascertained, we returned to each raised beach to collect the second type of data. A small pit was dug into the uplifted beach face and a collection of shells was taken to be analyzed using radiocarbon dating techniques (the carbon-14 isotope) to reveal how long ago it was that the the organisms had died. This, in turn, established the approximate age of the beach. The two pieces of information, elevation and beach age, could then be used to plot an uplift curve that, in conjunction with numerous other curves around the country, showed how the various ice masses must have interrelated, at times coalescing together and at other times operating independently.

The work that John England and I did that summer fed into the future global understanding of climate change and the stability of shorelines. John never left his High Arctic research. He carried out field work for the next fifty years and is now a Professor Emeritus at the University of Alberta. In 2019, John was awarded the Order of

Canada for the significance of his contributions to Arctic geomorphology and climate science.

It is difficult to describe what a pleasure this work was. On top of the staggering beauty of the environment, we had the excitement of self-sufficiency. We were responsible for everything. Our little boat was packed with all the camping gear and food we needed plus a rifle and ammunition, a radio, and our surveying equipment. We were overloaded, in fact, and had to park our bodies wherever they could fit best, being careful not to damage any critical gear. It was entirely up to us to explore, find the raised beaches, and identify the marine limits. We soon became proficient nomads, voyaging from one bay to another, setting up our tiny, two-man tent, usually on a comfortably flat, sandy top of a raised beach, staying in touch with others by radio, and perfecting our survey and sampling skills. Our diet of freeze-dried delicacies was mercifully supplemented with Arctic char, which were caught easily in nearby streams and lakes. One lake yielded such abundant catches that we dubbed it Brutal Lake.

Toward the end of the field season, we were pulling into a small bay when John exclaimed, "Patrick! A polar bear!" Sure enough, we were privileged to watch a snow-white bear haul itself out of the water and onto the beach. He immediately sat down on his haunches (based on the size of the footprints, *he* is probably right) and stared hard at us, watching our every move. There was nothing friendly in his examination. We idled the engine while we watched him back and wondered what he might do next. John was excited. He had already had a few Arctic field seasons, but this was his first polar bear sighting. I had heard him lamenting his bad luck before and realized the importance of this sighting for him. Everything we had heard about encounters with polar bears had made it clear that they could be extremely dangerous and should be treated with great caution. After a while, our friend stretched himself, gave us a final steely glance, and ambled off up the valley we had come to survey.

The tide was low, and we could manhandle the heavily loaded boat only a few feet onto the intertidal flat. We attached it with a long rope to a boulder on the shoreline. Since we had the rifle, there seemed no reason to think the danger was excessive as long as we kept a sharp lookout. The bear was now a small white blob moving away from us far up the valley. Just for fun, I measured the prints that were perfectly formed on the sandy beach. The hind paw was eleven inches wide and seven and a half long.

Our view of the bear was soon obscured when a dense fog rolled in, quickly followed by harsh sleeting rain. What was it about that day? Breakfast had been a failure. The bannock I had made from Tea-Bisk had stuck to the pan in a horrible mess. Held together with peanut butter, it had seemed enough to line our bellies, but for some reason, we hadn't bothered about lunch, so our blood sugar was probably low. And now, the day that had started out so fine had become cold, wet, and cheerless. I was at the theodolite, but John had taken the rod too far for me see its markings. The rain was mucking up my glasses and the theodolite lens. I waved my arms in the signal to move closer, but John appeared to be waiting for me to move the theodolite. Annoyed now, I left the instrument and went to straighten him out in person, only to find that he was pissed off at me. He claimed I had signalled that the measurement was taken. Fuming, I marched back down again, finally got a measurement, and we continued with the survey.

Except that the same kind of miscommunication was repeated... and repeated again. Our normal good cheer and witty repartee vanished as our tempers reached breaking point. We made it to the marine limit in a hopelessly foul mood. The weather had socked in completely. Any chance of seeing our bear, should he decide we were on the menu after all, was a non-starter. We loaded our gear and stomped in grim silence back down the valley to the boat, then stood on the beach and stared down at it in shock and disbelief.

We still had a lot to learn about seamanship. We had left our craft unprotected on a lee shore. While we were surveying, the wind had got up, the tide had come in, and waves were now breaking over the transom. The boat was completely swamped.

We could see all our gear wallowing in the rough water that filled the boat up to its gunnels. The radio, the Coleman stove, our packaged food, our tent, sleeping bags, and clothing were all underwater. The labels of the freeze-dried, tinned food floated on the surface. And the toilet paper? GOOD HEAVENS! That didn't bear thinking about.

In a rage, we unloaded everything onto the beach and pulled the boat high above the tide line. What was there to do? Absolutely nothing except carry on as though everything was normal. Without a word, we set up the dripping-wet tent. It was pouring rain anyway, so what difference did it make? Now it would be wet inside as well as out. It did not go unnoticed that our chosen campsite was right beside the polar bear tracks. Wringing out the foam mattresses and sleeping bags, we laid them inside. The radio and Coleman stove were both toast. We didn't even try to revive them.

Then John demonstrated his true leadership qualities. Scrounging around in the supplies, he found cans of peanuts and jars of trail mix. Reaching into his packsack, he produced a bottle of Ballantyne Scotch Whisky! Together, we huddled in the tent and opened the snacks. Together, we passed the bottle back and forth, forth and back. Together, we felt the woes of the world slip away. And together, we forgot the day's blunders and misunderstandings. Our habitual good humour returned, wittier and funnier than ever before. Our friendship was completely restored. And the polar bear? As the whisky went down, our conviction rose that every real or imagined noise from outside was the bear returning. But there was nothing to worry about. We took it in turns to rush outside the tent, waving the rifle and shouting our lungs out. "Well done, John," I congratulated him. "You sure must have scared him off that time."

"Actually, Patrick, it must have been you last time because I don't think he was there now." Whatever the case, our efforts must have worked. We never saw him again.

∞

You may think that wet sleeping bags would have been enough to keep us awake all night. Quite the contrary. Still fully clothed, we got inside them and slept like babies. I awoke with a shaft of sunshine playing over my eyelids. It was coming in through the tied door panels of the tent, which were flapping gently in a light breeze. I lay there marvelling at how the heat of your body, which I knew could dry wet clothes if put inside your sleeping bag the night before, could also dry the sleeping bag itself. It was at least a lot less wet than when I had climbed into it. John was still snoring peacefully. I felt calm and serene. I knew without having to look outside that it was a beautiful day.

Slowly, I became aware of some peculiar noises. A pitter-pattering of some kind. I listened hard but couldn't figure them out. They were irregular and sporadic, three or four together, silence, then another group. They sounded close, as if coming from the tent itself. I strained my ears. Yes, a distinct pitter-patter, pitter-patter, light as a feather. Then a gentle whooshing noise followed by a soft *plop*. I stuck my head out.

"John…John! Wake up! You've got see this!" I whispered and prodded the prone form with my foot. Still groggy, but intrigued by the urgency in my voice, John pushed the tent flap further to one side and stuck his head out beside mine. First, he noticed the loveliness of the morning, made even lovelier by the misery of the previous day. Then, he spied the lemmings. A line of them (the technical term for a group of lemmings is a slice) was congregated by the main guy rope that led to the ridge pole that held up the tent. Each was waiting its turn to climb up the guy rope, which they could do with some agility. They were beautiful little creatures about six centimetres long and covered in rich brown and rust-coloured fur. Reaching the ridge pole, they scurried along it (pitter-patter, pitter-patter) and then... this will be difficult to believe, they launched themselves onto the sloping fabric of the tent and slid down its length, landing on the ground with that soft plop I had been hearing. Each one got up with a little shake and hurried back around the tent to get back in line for another go.

This spectacle was too much for us. Holding back laughter so as not to frighten them, we collapsed back onto our sleeping bags and stared up at the roof of the tent listening to the pitter-patter of tiny lemming feet and watching the shadows on the sloping wall of the tent as the lemmings whooshed down its side. It was irresistible. I don't know who tried this first, but we found that a quick, light whack on the fabric in the middle of the slide sent the lemming somersaulting through the air to land several metres away from the tent. We took it in turns, whacking while the other watched. Was this cruel? On the contrary, the lemmings loved this new addition to their sport. Dusting themselves off, they couldn't wait to get back in line.

The next few days were not nearly as entertaining. We knew that it would eventually dawn on someone that we were no longer making our regular radio skids. Then Dave would be sent out to find us, but until that happened, we had to make do. This meant subsisting on whatever we could salvage from our flooded food supplies. We had some canned tuna, which we ate with applesauce (not bad). Some of the Tea-Bisk was still dry, but we would have to be desperate to try eating that raw. We kept working on our surveying and sampling as best we could. The weather was bad, with unpredictable winds, and the fjord was often dangerously choppy. More than once, we had to turn tail and quickly head back to where we had set up camp. Finally, four days after our disaster, the sun came up. We decided to take a break and spend the day scrounging as much driftwood as we could find on the beach. There wasn't much, but by the end of the day, we had a decent pile. We had spent part of the morning randomly opening cans of freeze-dried food. With the labels gone, there was a certain excitement in seeing what treasures we might uncover. By evening, a mess of pork chops had been soaking in water most of the day. Peas, mushrooms, and instant mashed potatoes were also ready for cooking.

We built a roaring fire and took our time getting the coals just right for cooking. My diary records that we ate nine pork chops each, complete with as many of the edible vegetables as we wanted. Bloated, we retired to our sleeping bags. And at that point, Dave arrived in the helicopter. It had finally been decided that he had better come looking for us.

The summer drew to a close, not with a whimper but in a mad scramble to get the Generator and Inugsuin camps packed up and put away to survive the upcoming winter. Our Fox-Charlie camp also needed to be taken down. The weather, never good for long, got steadily worse. The perpetual winter night was on its way, a few minutes of darkness at first, then becoming progressively longer as each day passed. I flew back and forth from one camp to another with Dave, enduring more than a few exciting flights. En route from Inugsuin to Fox-Charlie one time, the helicopter was tossed around in weather so rough that I was sure the end was near. The expression on Dave's normally unreadable face did little to reassure me. The snow beat hard against the bubble, and in one mountain pass, the strength of the wind forced us to turn back and find another route.

The Generator Lake camp, now that the snow had gone, looked as messy and ugly as Fox-Charlie. I was dismayed. We had been just as bad as the military. Worse really, since we should have known better. It reinforced my growing opinion that we humans really care little for the other inhabitants of our planet. No doubt the wandering wolves and caribou probably weren't worried about the aesthetics of this small area of spoiled land nestled against the ice cap. But what about all the plastic debris, half-empty oil containers, and endless steel drums? What about the smaller creatures and the food chain? It has always seemed like a no-brainer to me that such destruction cannot possibly continue without consequences to ourselves. I worked for some days, cleaning up the site as best I could.

As the season neared its end, the work became frantic. There seemed to be an almost desperate excitement to return south. My feelings were mixed. I was excited too, but also worried that after four months in the isolated, otherworldly Arctic I would feel estranged from town life and even from my friends. Would our paths already have separated too far? I had given little thought over the summer to The Plan I had made with Don Hindle. In fact, I had spent a significant amount of time talking to Dave about how to become a helicopter pilot (it turned out my eyesight would preclude this career option), but I was also still attracted by the idea of travelling in South America and had spent a fair bit of time studying my Spanish book. I knew one thing for sure. I wasn't going back to university.

As Dave and I were flying back to Inugsuin in the twilight, we heard voices on the radio that seemed to be talking about *a wolf in the fjord*. We looked at each other in puzzlement. A wolf in the fjord? What could that mean? We soon found out. As we flew over the fjord, we spotted a small, red-hulled ship, the *Wolfe*, sitting at anchor offshore from our camp. It was the sea lift bringing in supplies for next year's field season. The beach was already full of people unloading a landing barge. Dave flew low and circled the ship. We saluted by rocking the helicopter from side to side. I blew the whistle on my life jacket as loud as I could. At the camp, we found that Butch and Salomonee had recently been through and brought us caribou. I made a caribou stew and Dave baked a cake. The *Wolfe*'s captain and chief engineer showed up with fried chicken and a bottle of gin. A magic show followed to great applause. We were invited for breakfast on board before they left, but the combination of hangovers and their 6:00 a.m. departure deadline put an end to that.

August 23 was our final day at Inugsuin. We were up at 6:30 to make the final caches, get the roof tied down, and finish up a surprising number of other jobs that demanded attention. Dave flew to and from Fox-3 all day, ferrying us out. I was the last to leave. He picked me up in the mid-afternoon and gave me one of his best rides ever, showing off some of his acrobatic tricks. Like his predecessor, Jim-the-pilot in Operation Bow-Athabasca, he capped the performance by flying backwards. At Fox-3, I had to settle for a cold shower. All the hot water had been used up by the others when my turn came. Our last supper was unusually good, the movie predictably terrible. To top off the evening, a magic show was called for and a good time was had by all.

Then the Arctic reminded us of what it liked to do best. I can't remember how many times this happened to me in my fieldwork days. All packed up and nowhere to go. A DC-3 was meant to arrive first thing in the morning to take us to Frobisher. Instead, we woke to dense fog, high winds, and zero-zero visibility. There would be no DC-3 that day. Or the next. On the following day, August 25, the fog lifted, but winds were gusting to seventy kilometres per hour. No plane again.

Those three days were excruciatingly slow. There was nothing to do but wander around, eat too much, watch bad movies, and read. Finally, on the 26th, lo and behold—a clear sunny sky and minimal wind. We were frantic with anticipation, and when the DC-3 arrived at noon, we decided to charter it all the way to Ottawa. Fine weather or not, we still had a bumpy ride to Frobisher, and while the plane refuelled there, we walked up to the Federal Building to have Sunday supper. Taking off again at 7:00 p.m., we headed to Schefferville, Quebec, almost exactly a thousand kilometers south. A long, dreary flight, then wonderful to arrive in the middle of a clear black night with stars and shifting northern lights overhead. We paced the runway in the dark while another refuelling took place and were soon in the air again for another twelve hundred kilometres south-southwest to Ottawa. Lost in thought as we lumbered through the night, I realized with startlingly clarity what I really wanted to do. Whatever happened, I would get hold of Don Hindle as soon as possible, and we would carry out the third part of The Plan. We touched down in fog and pouring rain just in time for breakfast in Ottawa.

REFERENCES

Montgomery, Rutherford, G: *Carcajou the Wolverine.* Caxton Press, 1936

Morris, Jan: *Fifty years of Europe: An Album.* Random House Publishing Group, 1997

Slocum, Joshua: *Sailing Alone Around the World.* The Century Company, 1900

Smeeton, Miles: *The Sea was our Village.* Gray's Publishing Limited, 1973 *Once was Enough.* Rupert Hart-Davis, 1960

Tilman, Bill: *Ice With Everything.* Gray's Publishing Limited, 1974

Walker, Dora: *Freemen of the Sea.* A. Brown, 1951 *They Labour Mightily.* A. Brown and Son, 1947

ABOUT THE AUTHOR

Patrick McLaren is a well-known geologist who was a Research Scientist with the Geological Survey of Canada before founding a consulting company in the United Kingdom. As a consultant, he lived on his sailboat, travelling the west coast of North America and collecting sediment samples for environmental research. During his career, he became concerned about the fate of the planet, and he has fought against fraudulent science being applied for economic gain by research sponsors. In his field, Dr. McLaren has published chapters and written extensively for scientific journals.

The Unlikely Adventures of a Geologist: Magic Travels is Dr. McLaren's only writing excursion outside of science and the first of three volumes about his life. He lives with his partner, Susan, in Victoria, British Columbia, Canada.

Printed in Canada